FE ELECTRICAL AND COMPUTER EXAM PREP

The Ultimate Study Guide with 1500+ Practice Problems, Detailed Solutions, In-Depth Reviews, and 4 Mock Exams to Pass with Ease & Achieve EIT Status

By

LEARNOVA

ISBN: 978-1-80266-996-1

First Edition

Table of Contents

Welcome to Your FE Exam Prep Journey

Preparing for the FE Electrical and Computer exam demands a structured and comprehensive approach. Our research found that the resources currently on the market do not meet all the criteria to fully prepare candidates for this exam. Their focus is usually on just one aspect of exam preparation, such as only theoretical reviews, or only practice problems, or only mock exams. Additionally, many of these resources have not been updated for years, so they don't reflect the latest exam specifications.

Here at **Learnova**, we recognized the need for a unified solution. Our goal was to create what you have in front of you now: the ultimate, all-in-one resource with everything you need to pass the exam on your first attempt. This unique study guide allows you to concentrate on what truly matters without wasting time searching for the right materials. Written in simple language and an easy-to-follow style, it provides a comprehensive package that includes theoretical reviews, practice problems, and mock exams, all tailored to the latest exam requirements. Additionally, it provides expert advice on how to prepare mentally, psychologically, and physically for the exam.

You'll learn how to:

- Plan and follow an effective study schedule
- Gain practical mastery over the concepts and formulas

- Reduce your stress levels before and during the exam

Intended for final-year students, recent graduates, and seasoned professionals returning to study, this study guide supports you every step of the way, ensuring you are as prepared psychologically as you are intellectually.

Maximizing Your Use Of This Study Guide

The first five chapters are not subject-specific. They cover:

1. An overview of the FE exam structure and scoring, and the value of professional certification.
2. Familiarizing yourself with the Reference Handbook with navigation techniques and advanced tips and tricks.
3. Building an effective study plan, creating a study hours log table, tips for efficient learning, time management strategies, and proven techniques to overcome test anxiety.
4. The best of the approved calculators and advanced calculator tips and tricks.
5. Last-minute review strategies, essential exam-day checklist, and time management during the exam.

These chapters will not take you long to read, but they are as important as the chapters on the knowledge areas, so do not skimp on them. If you want to be in the best place — mentally, psychologically, and physically — to focus purely on answering the exam questions, then you need to:

- know the exam structure
- understand how to get the best out of the Reference Handbook
- understand, and know how to use the allowed calculators
- create an achievable and practical study plan
- overcome test anxiety

Chapters 6 to 18 cover each of the 17 Knowledge Areas (topics) that the exam expects you to know and understand. And rest assured, this study guide includes only the topics you need to study for the current exam. No extra topics were added, as all they do is add to your load, your uncertainty, and possibly your stress levels. We followed a "Keep it Simple" strategy when we designed and prepared this guide.

At the end of each of these review chapters, you will find the following (in this order):

Practice Problems: These are online, easy-to-navigate and printable practice problems available for download.

- You will receive immediate access to 1000+ up-to-date exam-like practice problems with step-by-step solutions. (Note: We did not include them in the physical book for 3 reasons: 1. To allow us to catch small mistakes and correct them without having to wait for the next edition; 2. It allows us to update them to the latest exam specifications every time the curriculum changes; 3. To avoid overwhelming you with a book of over 800 pages, which can be intimidating and less practical for study purposes.)
- Two weeks after the initial download, you will receive a full-length ***practice exam*** with step-by-step solutions. You will get one new practice exam every week for 4 weeks to simulate the exam day. (We don't include them in the guide to allow you to get to know the book and not be overwhelmed by the volume of practice problems until you're ready for them.)

Video Resource Recommendations: We suggest some highly recommended video resources to further your studies and/or for clarification of complex topics. We chose not to create and sell our own video resources because the most detailed and well-explained resources you need to pass the exam are already available online for free. This book is the only product we'll ever sell because our primary interest is your success in passing the exam, not in profiting from sales of supplements.

Fillable Tables: Visual aids to help you understand and remember the key concepts and formula names you'll encounter in the exam. We intentionally exclude the formulas themselves – providing only the concept and formula name – to ensure you use the FE Reference Handbook and memorize their locations. Our primary goal is to ensure you pass this exam, and your goal is to pass this exam, not own a fancy book, right? While not all highlighted concepts from the tables may be in the Handbook, we believe these are crucial for passing the exam and we cover them through the practice problems. We also realize the importance of having the key formulas written down, so we added a feature for you to do just that (see next bullet point.)

Notes/Equations Summary Blank Sheets: We deliberately include blank pages for you to write down important notes and personal summaries of equations.

Diagnostic Test Checklists: These help you identify knowledge areas where you need to improve and track the tasks and actions that still need your attention.

Reporting Errors and Updates

This study guide has been meticulously reviewed and refined through multiple cycles to ensure high quality and readability. Despite our best efforts, some errors may have slipped through. We welcome your feedback on any errors, issues accessing the online resources, suggestions for improvements, or any questions you may have. Please feel free to contact us at info@learnovapublishing.com

We are committed to responding to all emails within 24 hours, seven days a week.

About Learnova

Learnova stands for innovation in learning. We are dedicated to creating modern, up-to-date, and practical study tools that cater to the needs of today's students. By collaborating with the smartest minds in the field, we ensure that our content is accurate, relevant, and effective.

Our mission is to empower you to achieve your career and exam goals. We do this by developing unique and innovative products that are easy to use both in print and online, on any device. Learnova products are designed to measure your progress and mastery of skills, providing you with a comprehensive understanding of the subject matter and the confidence to pass your test.

CHAPTER 1

Navigating The FE Electrical and Computer Exam

1.1 Overview of The Fe Exam

The FE Electrical and Computer exam is a computer-based test (CBT). Candidates effectively have 5 hours and 20 minutes to complete 110 multiple-choice questions. This means that, on average, you have two minutes 54 seconds per question. This exam is a crucial step in becoming a licensed professional engineer. It is designed for engineers in training and students who are in the final year of their undergraduate electrical engineering degree. The FE exam is administered all year-round at NCEES®-approved centers, and annually at centers with a smaller number of examinees.

The all-year-round exam centers use what the NCEES® calls a "linear-on-the-fly algorithm" to determine the questions. This means that all examinees for any particular exam have the same number of questions on the same topics, but no examinees will have the same set of questions. The algorithm ensures a unique exam for every sitting. The questions will be at the same difficulty level relative to any other sitting.

At the smaller centers, where exams are administered on a single day each year, a different testing model is used. In these cases, all exam candidates will get the same set of questions. The types of questions, and the formats in which they are presented, are the same at the annual events and the daily testing centers, irrespective of the statistical model used. In all cases, a $225 exam fee is payable directly to NCEES®.

1.2 Value of Fe Electrical And Computer Certification

The importance of this exam in the field of engineering, particularly its role in the licensure process, cannot be stressed enough. Licensure can help engineers progress in their careers, while at the same time providing the public with assurance that engineers have a high level of educational qualification, work experience, competence, and professional accreditation. Licensure holds engineers to high standards of public health and safety. The criteria for licensure vary from state to state, but there are three steps that are common for most states:

1. Licensure requires professional engineers to have a bachelor's degree from a duly accredited university program.
2. Most states require four years of work experience.*
3. Generally, licensure candidates need to pass both the FE exam and the PE (Professional Engineers) exam.*

*Some states may have other licensure requirements to meet. You need to confirm the requirements in the state or territory where you will be working.

1.3 Exam Structure And Scoring

The FE exam consists of 110 questions. The exam duration is six hours, and the time is allocated as follows:

- Two minutes to read and sign a simple confidentiality agreement
- Eight minutes for a brief tutorial
- Five hours and 20 minutes for the actual exam
- 25 minutes scheduled for a break.

The exam consists of multiple-choice questions, where you select one answer from a choice of up to four options, and an unknown number of Alternative Item Type (AIT) questions. AITs comprise four alternative ways of testing the breadth and depth of an examinee's knowledge. The AITs you will probably encounter are:

- Drag and drop—requires examinees to drag items to change the hierarchy, label an image, or match different items.
- Fill in the blank—The examinee has to complete the question by entering the missing information.
- Multiple correct—expects the examinee to select more than one answer.
- Point and click—the examinee needs to click on one or more parts of an image to answer.

All answers are either correct or incorrect. Partial answers are not scored, and there is no penalty for wrong answers. Your score is adjusted for the difficulty level and for small differences across the exam formats. Your ability level is compared to an average ability level for that exam. All the exams for a specific discipline are scored the same way, and no examinee, whether this is their first time or a repeat exam, is graded differently.

The test is divided into two sessions, with 55 questions being presented in each session. Between the two test segments is an optional 25 minute break. Not taking the break, or not using the full 25 minutes to refresh yourself or have something to eat, does not mean that time is added to your time to answer questions. Every examinee has the same amount of time, five hours and twenty minutes, to answer the 110 questions.

Before you finish any session, you are still able to review answers you were not sure about, and/or complete questions you skipped. But be aware of how much time has already elapsed of the 5 hours 20 minutes allowed for solving problems, so you are not left short of time to answer the remaining 55 problems. You need to find a balance here between your review and how much time you have left, because once you close the first session and submit your answers, you will no longer have access to those questions.

The number of questions per topic will follow the outline provided by NCEES®, and discussed in Chapter 3. However, please note the topics will not be labeled on the test.

Exam results for CBTs will be emailed to you by NCEES®, generally a week to 10 days after the exam. You will also receive instructions on how to view your results in your MyNCEES® account. Results are given as a pass or a fail, not as a percentage. The notification will include information specific to your licensing board, and how you should proceed based on your results.

CHAPTER 2

Mastering The NCEES® FE Reference Handbook

2.1 Introduction To The Fe Reference Handbook

The importance of the NCEES® Reference Handbook cannot be stressed enough. Apart from a calculator, your retained knowledge from your degree studies and any revision you have done, the Reference Handbook is your only source of help. It is in digital format, and the version for your use during the exam excludes the index.

We **HIGHLY** recommend that you download it from your myNCEES® account and keep it always open on your computer when reading this study guide and using our resources. It is important to start using this reference book as soon as possible, because the better you know it, and where the specific topics that you need to study are, the quicker you will be able to navigate through it during the exam.

The Reference Handbook is about 500 pages long, but it contains concepts and formulas for seven different types of FE engineering exams, including Chemical, Mechanical, and Civil. The topics that are relevant to the FE Electrical and Computer exam are the only ones that you need to focus on. They are listed in **Chapter 3** of this study guide along with the minimum and maximum number of questions that you can expect for that topic. We also look at how you might benefit from knowing which topics carry a higher number of questions so you can be better prepared for them.

To be able to use the NCEES®-provided Reference Handbook efficiently under exam conditions, you need to know the location of each relevant chapter and the concepts it covers. It's a good idea to make notes in your study guide about the relevant page numbers and concepts covered. We have included fillable tables throughout the study guide specifically for this purpose. Writing something down helps you to memorize it. If you ever wrote crib notes during your school days, and then deliberately left them in your room, you will know the power of recall that writing things down gives you. You remember them. Knowing the page numbers of concepts and formulas will reduce the time you spend searching through the Reference Handbook during the exam.

2.2 Familiarizing Yourself With The Reference Handbook

The Reference Handbook is a goldmine of info for the exam, and you should familiarize yourself with where to find the tables, formulas, charts, and other similar vital information it provides. You need to have an excellent grasp of where to find the relevant information quickly so invest a significant amount of time in browsing it. It will pay dividends during the exam.

Study from the pdf version of the Reference Handbook, rather than the hard copy, so you become familiar with using the Ctrl+F search function. The Reference Handbook has many chapters that cover different knowledge areas, but you will not find chapters on Linear systems, or on Circuit Analysis as these are covered in Electrical and Computer Engineering. This is just one reason that you need to work with the Reference Handbook while studying.

2.3 Quick Navigation Techniques

The CTRL-F search function can be painful, particularly if it finds multiple occurrences of your search word or phrase. Knowing how to use the advanced search options, such as matching case, or whole words only, is vital to improving your search efficiency.

Take particular note of where in the Reference Handbook you will find the relevant tables and charts to help speed up your information retrieval.

Knowing the specific discipline in which the formula or equation will be found can speed up the search, but the first prize is remembering the page numbers of all the information vital to you. Excellent recall of memorized material, such as page numbers, is vital.

Many years ago, a Dale Carnegie course taught a simple way to memorize a list of 10 items. These 10 items could be page numbers, for example. The trick is to make it visual, as our brain works better with images than it does with words. Associate a different image with each number, from 1 to 10, based on a rhyming pattern. So, for example:

1. Run	2. Zoo	3. Tree	4. Door	5. Hive
6. Tricks	7. Heaven	8. Gate	9. Time	10. Hen.

Next, you need to tag the items to be remembered to one of the visual images you have in your head. For example, if Run is a horse race, put what you want to remember in the saddle of the leading horse. For Zoo, picture the monkey house, and they are throwing whatever you want to remember to each other, or at you. Your Tree is a Christmas tree and what you want to remember is the top decoration of the tree. Or what you want to remember is being hatched under the Hen. Make the images outrageous, fun, and memorable. Next time you need a shopping list or a to-do list, start using this method. You'll be amazed at how quickly and easily you start remembering things. By the time you walk into the exam, your memory will be stronger and recall will be easier.

2.4 Understanding Units

The F.E. Exam uses units from both the International System of Units (SI) and the United States Customary System Units (USCS). The equations in the Reference Handbook are non-specific, in that they do not distinguish between SI or USCS units. So the onus is on you to pay close attention to the possibility that USCS units may be used in a problem where the solution must be in SI units. Double-check yourself, particularly with problems that use USCS units, to make sure you have done any necessary conversions and that the units cancel out to the unit of your final answer.

2.5 Practicing With Reference Handbook-Based Questions

We recognize that students need practice problems covering all exam topics, with solutions, to fully understand the different concepts. This is why we have made 1500 practice problems, with step-by-step solutions, available. As stated earlier, these are printable, easy-to-navigate, and available for download. More importantly, a set of practice problems and solutions is available for every one of the exam topics. This is a unique feature of this study guide. **NOTE:** The link/QR code to access the practice problems is at the end of each of the review chapters (chapter 6 onwards). We strongly recommend that you work through these methodically.

2.6 Advanced Tips And Tricks

We include several memory improvement techniques below. These are sorted alphabetically, as we cannot recommend any one method over the next. It is an individual choice as to what works best for you. They are all tested and trusted ways of improving memory performance and will help you memorize the material. The earlier you start using them, the better, as it will take a few weeks for the results to start showing.

Active recall: Regularly test yourself on the material you've learned. For instance, after studying a section of the NCEES® FE Reference Handbook, close the laptop and write down everything you remember. This exercise forces your brain to recall information actively, which strengthens your memory.

Association: Relate like information to other like information. Similar to Chunking, putting bits of information in groups, particularly if it doesn't have to be in order, makes it easier to remember.

Chunking: Break up information to recall into smaller parts. For example, instead of trying to memorize a long equation in one go, break it down into its individual components, look for patterns, and understand each part before putting it all together.

Memory palace: To create a memory palace, start with a place you know well, such as a family home, and using your imagination, draw a full-color 3D mental map of the palace. Next, identify specific rooms/spaces and put the individual items you want to memorize in there. The more vivid, silly, or outrageous the placement, the better. Your brain finds it easy to recall new things that are linked to familiar things.

Mnemonics: Using techniques such as rhymes, acronyms, or visual imagery to remember things is a great way to make memorization easy. Your geography/art teacher might have taught you a mnemonic to memorize the colors of the rainbow in the right order. ROYGBIV becomes "Right On Your Game, Best In View," or "Richard Of York Goes Back In Vain."

Organization: Grouping related items together, such as drawing mind maps, makes them easier to remember. This technique is really useful for complex information, particularly if it must be chronological. It helps you remember information in a specific order.

Recitation: Read your study material out loud. Your brain will love it. It's great if you can do this with a friend, but not essential. Have you ever wondered how actors can remember long passages of dialog? They read aloud to each other during rehearsals. Importantly, recitation helps you remember information in the original sequence.

Spaced repetition: Repeat the material you need to remember at spaced intervals of time. Start with daily repetition for a week or so, then increase the interval to weekly repetitions, then monthly. This method works best if you have given yourself long enough to do this with all the material you need for this exam. This is an antidote to fading memory for both short- and long-term retention.

Visualization: Create vivid, full-color mental images of the study material. For example, visualize how an electrical circuit works by imagining the flow of current through the components. Associating abstract concepts with visual images makes them easier to remember.

CHAPTER 3

Your Exam Preparation Blueprint

3.1 Building An Effective Study Plan

If you're taking the FE exam at the end of your final year, a lot of the information will still be fresh in your mind. Then, at a push, you might be able to study for the exam in two weeks. But there is a big IF involved here. It's only *if* you still retain a lot of the information from your studies. If there is a faint possibility that this is not the case, then don't take a chance; but rather plan and draw up a suitable study plan as set out below.

1. Make a list of all the concepts/categories in which you need more practice.
2. Ensure that you master the concepts before doing the exercises.
3. Understand the exam's breadth and depth of content, and give more time to the concepts that still need intensive work in your personalized study plan.

Only you know your engineering strengths and weaknesses, and while you already plan to focus on the areas you have identified, do not ignore the rest of the curriculum. There is no harm in reinforcing the known and understood material - it helps to keep it fresh.

3.2 Create A Study Hours Log Table

This table will encourage you to develop a regular and workable study routine. You decide the number of hours per day that you study, and whether you study every day of the week. Because it's your study plan, designed by you, it should be easier to stick to it. What is important though, is to accumulate a significant number of study hours before you take the exam. Your study hours plan will help you to do that.

When you set up a study schedule, aim to achieve, or get close to, 150-300 hours of completed study time before sitting for the exam. This equates to 2/3 hours per day for 2/3 months. One of the criteria that needs to be considered is what your personal circumstances are in terms of available time for studying, and how long it has been since you finished your degree program. If it is anything more than six months since you last studied, you will have to invest a lot more time in your studies.

No need for panic, though. We designed this study guide to be flexible, allowing you to concentrate on the areas where you need the most support. Whether you have enough spare time to study or less time than you'd like—Learnova will help you make the most of it.

Knowledge Area	Session 1 Date - Time	Session 2 Date - Time	Session 3 Date - Time	Session 4 Date - Time	Cumulative Time
Mathematics					
Probability and Statistics					
Ethics and Professional Practice					
Engineering Economics					
Properties of Electrical Materials					
Circuit Analysis (DC & AC Steady State)					
Linear Systems					
Signal Processing					
Electronics					
Power Systems					
Electromagnetics					
Control Systems					
Communications					
Computer Networks					
Digital Systems					
Computer Systems					
Software Engineering					

3.3 Efficient Learning

You must become as familiar with the NCEES® provided Reference Handbook as you are with your hands. Knowing the location of each relevant chapter in the Reference Handbook along with its concepts, will save you some much needed time. Ideally, you need to master the concepts in each topic, or at the least, understand them, before doing the exercises. This table sets out the topics you need to know and understand for the FE Electrical and Computer exam, with the minimum and maximum questions you can expect for each topic.

NOTE: This table shows the current min/max questions, but they often change, so we suggest you refer to the official exam specifications on the NCEES® website.

Knowledge Area / Topic	Min/max questions
Mathematics	11/17
Probability and Statistics	4/6
Ethics and Professional Practice	4/6
Engineering Economics	5/8
Properties of Electrical Materials	4/6
Circuit Analysis (DC & AC Steady State)	11/17
Linear Systems	5/8
Signal Processing	5/8
Electronics	7/11
Power Systems	8/12
Electromagnetics	4/6
Control Systems	6/9
Communications	5/8
Computer Networks	4/6
Digital Systems	8/12
Computer Systems	5/8
Software Engineering	4/6

You will notice that **Mathematics** and **Circuit Analysis** (DC and AC Steady State), have a minimum of 11 questions each and a maximum of 17 questions each. Along with **Power Systems** and **Digital Systems** (min. of 8 questions and max. of 12 questions each), these four topics potentially have half of the questions you will have to answer. Yes, that says four out of 17 topics will be worth almost 50% of the available marks for the exam.

Add in the next highest topic, **Electronics** at 7 min/11 max, and five topics will cover a minimum of 45 questions and a maximum of 69 questions.

This being the case, if some or all of these topics are not where you shine, you need to focus more of your learning time on them, because they have a potentially higher payoff in terms of the bulk of the questions you will face.

3.4 Time Management Strategies

We have already covered the importance of knowing how to get the most out of the CTRL-F search function in the section on **Quick Navigation Techniques**. The computers provided at the testing facility will be poor in processing power, (although they will have 24" monitors) so each word search could take a significant amount of time. Another quick navigation technique, the quickest by far, is knowing the page number. On the extreme right of the pdf, hover over the page number, and an editable block appears immediately to the left of it. Put in the page number, and enter.

Take some time to learn how to spot keywords that you can easily search for in the Handbook PDF. Note that single words are better than a phrase.

Making a to-do list is a great start, but it can't just be a list that you read once and then ignore for a week. It has to be an active exercise that you revise as priorities shift. It also has to be balanced and include time out for healthy eating, social activities, and exercise.

Three things that must be considered when drawing up a study plan are:

1. What is the longest period you can maintain your focus? This is important because your study plan must be built on manageable periods of time. If your attention span is short, say 25 minutes, then the Pomodoro technique might work for you (see below.)
2. You need to take a break before returning to class material. The question is, how long a break do you need? If you plan to study for about three hours per day, would this include breaks, or will the breaks be over and above the actual study time? If it excludes breaks, you could break it into four 45-minute periods, with three breaks of 20 minutes each, a total of four hours out of your day. You could shorten the first and last break by five minutes each, add those 10 minutes to the middle break, and use it to do 30 minutes of exercise. You could go for a run, do yoga, or tai chi, or go for a swim.
3. In what part of the day are you at your best? For some, it's in the morning. For others, it might be late afternoon/early evening. If you know what it is, factor it into your study plan.

These final few tips are more than good time-management strategies. They also include important lifestyle choices for general health and happiness.

- **Remove any distractions, such as your phone**. Put it on silent and hide it under your pillow on your bed.
- **Set priorities and reward achievements**. Consider rewarding yourself for ticking off to-do list items. Many small achievements add up to big progress, so give yourself small rewards, such as an extra chocolate, or an earlier coffee break.
- **Stick to a healthy diet** with plenty of fresh fruit and vegetables. Nuts and eggs are good food for the brain, so include them.

- **Exercise daily. This improves blood flow**, eases stress, and releases endorphins to make you feel good about what you've done.

- **Try the Pomodoro technique**, in which you work for 25 minutes, then take a 5 minute break. This works well to improve your focus and concentration. If this technique is new to you, try it for a few weeks and see if it improves your productivity.

- **You have 2-3 minutes to solve any given problem** on the exam, so time yourself when doing the exercises. To help you to do this, here are two little-known tips:

    - Most questions on the FE exam require a few steps to solve. This could include using the Reference Handbook to find an equation, or data from a table or chart. With each question, make an effort to reduce the number of times that you need to search for something in the Reference Handbook.

    - Most problems on the FE Exam can be solved in two steps using the most efficient method, so look for the two-step solutions when solving problems during your preparation. This guarantees you'll find those solutions during the official exam as well.

- **Tackle the trickiest task first**. This strategy is based on the idea that getting the most difficult task done makes all the others easier. It works best for morning people, but it still might work for evening people as well. You won't know until you try it for a couple of weeks.

3.5 Overcome Test Anxiety With Stress Reduction Techniques

Knowing how to manage stress and anxiety during your preparation and leading up to your exam, is key. Here are a few techniques worth incorporating into your plans.

1. **Breathing**. Setting aside a few minutes every day to practice mindfulness is a great technique to calm you down. It eases your body's stress response and pulls you back to the present moment. One method we recommend is *qigong*. Pronounced "chi gong," this is a millennia-old Chinese method of measured movements, combined with conscious breathing, to optimize energy within the body. It is calming and relaxing, helping you to release anxieties and negative thought patterns.

2. **Eat well, sleep for eight hours, and exercise**. We have already covered these because they are vital for your health. For your body to give you its best, make sure you sleep at least eight hours, eat enough low-GI foods (or slow-release carbs), drink lots of water, and do at least 30 minutes of daily exercise.

3. **Set realistic goals**. Setting goals is never easy. Often, we overestimate our abilities or underestimate how long something will take, setting ourselves up for failure. Not being able to meet unrealistic goals discourages you, opens you up to negative thoughts, or upsets you. So be realistic on the generous side. Give yourself more time initially, and as you get better at reaching goals, you can adjust the scheduling.

4. **Study with somebody**. In 2004, a research paper established that revising with peers is a good technique to relieve stress and tension. It allows you to share your notes with a friend, or colleague, and hear them share their notes with you. This puts you in an alternating role of student/teacher.

5. **Pace yourself to prevent panic**. Panicking is not going to help you before or during your exam. If you experience any sign of panic, pause. Stop what you're doing. Stand up and turn your back on the task. Breathe out, dropping your

shoulders. Then breathe in deeply and slowly as you raise your arms above your head. Breathe out slowly as you bend to touch your toes. Do this six or seven times, have a glass of water with a sprinkle of salt, and return to your problem. Manage your stress by looking at your task with a view to breaking it down into smaller components.

6. **Believe in yourself**. Self-belief is crucial when facing any challenge. Visualizations are a great aid here, and so is reframing. So, instead of thinking, "If I don't ace this exam, I'll never get that job, that promotion." Reframe it to, "It'll be a magic moment when I get that job/promotion because I aced the exam." You're doing this for you, making yourself proud.

3.6 Making The Most Of Practice Exams

In our introductory chapter, we said "Two weeks after the initial download, you will receive a full-length practice exam with step-by-step solutions. You will get one new practice exam every week for 4 weeks to simulate the exam day."

These four mock exams are one of our study guide's biggest strengths. The benefits for you of practicing with mock exams are invaluable. For the best results, you need to simulate exam conditions, including timing yourself. The problems you need to solve for the mock exams will differ from the practice problems in that they are not ordered by topic so they more accurately reflect the exam conditions.

Before you start, prepare your snacks, drinks, and energy boosters as you would for the big day. Get a clock, separate from your watch (which you may not be allowed to take in the exam), and time how long it takes you to answer the first 55 questions, including your review of problems you flagged to go back to. Then, take a break of no more than 25 minutes.

Subtract the time you spent doing the first 55 problems from 5 hours 20 minutes. Then set an alarm for the remaining time, and do the last 55 problems. Don't despair if you don't get through all the last set of problems. This is your first exam simulation, and we have three others for you to try after every week of study. Of course, you don't have to do them weekly. If you have a long enough study schedule planned, you could do them every fortnight.

Take note of how much better you got with time management (or ease of finding the relevant information in the Reference Handbook) as you get more used to the mock exams.

NOTE: The links/QR codes to access the practice exams are at the end of each of the review chapters (chapter 6 onwards).

CHAPTER 4

Calculator Training Before the Exam

4.1 Approved Calculators

Casio, Hewlett Packard and Texas Instruments are the three calculator brands approved for use in the FE exam. The relevant model numbers are listed below. These are the only acceptable calculators for use in the 2024 exams:

- **Casio:** All fx-115 and fx-991 models (Any Casio calculator must have "fx-115" or "fx-991" in its model name.)
- **Hewlett Packard:** The HP 33s and HP 35s models, but no other HP models.
- **Texas Instruments:** All TI-30X and TI-36X models (Any TI calculator must have "TI-30X" or "TI-36X" in its model name.)

If you already have one of these calculator models and are comfortable using it, there is no compelling reason to change. At worst, have a spare one to keep in your locker and a spare battery - just in case. However, if your current calculator is not on the list of approved calculators, here are a few things to consider in your buying decision.

The HP35s model was discontinued in 2021 but is still available—new and used. While it is a programmable calculator and can store 27 programs, it doesn't (for example) have numerical derivatives or matrix operations as built-in functions. The essential question when it comes to the HP options is whether you want an RPN calculator or not.

The TI-36X's best feature, apart from being programmable, is its memory retention. Users can review previous calculations and results and copy a value for use in a current calculation. Its shortcomings are poor screen resolution, a table function that only allows one function to be entered, and matrices limited to 3x3.

The Casio calculators have apps accessible from menus, so you need fewer keystrokes for some functions. When you switch functions, the keys change their functions (like keys on a computer keyboard change function when you hold down the CTRL key.) You can enter two functions, not just one. On the downside, the Casio does not retain the memory contents when you switch between functions. It does not have rubberized feet, so it tends to shift on the table as you use it. (This could be solved with a bit of Prestik on the feet).

We suggest you use the Casio fx-115ES PLUS Scientific Calculator - Second Edition, simply because our research has shown that the Casio FX-115 series of calculators is the most popular type for NCEES® exams. All FX-115 models have been approved by NCEES®. However, it is a personal choice.

Your choice ultimately comes down to being familiar with your calculator and knowing how to use its features. So it's important to get it early and practice using it for the entire time you study.

4.2 Advanced Calculator Tips And Tricks

Knowing how to solve FE problems efficiently by using time-saving calculator methods is probably the best tip we can give you. It has been said that needing calculator tricks to save you time when getting through your FE exam, means you aren't ready for it. However, if you asked a roomful of engineers their opinion about this, you'd probably get as many answers as there are engineers. With that in mind, here is our first tip:

- As with the NCEES® Reference Handbook, you should always use your calculator when you are studying. That way you can be fast and comfortable with it on test day, and you can concentrate on thinking about the problems.

Our final tricks involve practice and more practice. Know your calculator well enough to do all of the following (plus other calculations that may come up):

- Solve for a variable in an equation without having to do the algebra by hand.
- Store values in variables: Many FE problems will take a few steps to solve. If you can, store a value in a variable, so you can use that value in another step. It's great if you can also recall the last calculated value, otherwise you'll have to type the value again.
- Solve a system of equations with matrices.
- Know how to cross multiply vectors and matrices.
- Solving equations that have 2 and 3 unknown variables.
- Solve quadratic equations.
- Calculate regression with mean and standard deviation.
- Solve integral and derivative calculations.

CHAPTER 5

Your Guide to Exam Day Success

5.1 Essential Exam Day Checklist

Exam day is here. By now, you should be fully prepared in terms of understanding and knowledge of the concepts related to each topic, have drawn up a checklist of items to take with you, and reviewed your strategies for managing time and stress during the exam. Here are some final tips to help you prepare:

1) Review Exam Rules: Read the rules on the official NCEES® website well before exam day. Then, review them again the day before to ensure you understand the exam length, optional break, and administrative processes*. This will help reduce stress on the day of the exam.

2) Arrive Early: Arrive at least 30 minutes before the exam to complete the check-in process without rushing.

3) Check-In Process at Pearson VUE:

- Agree to the NCEES® Exam Rules and Agreement with a digital signature.
- Present an acceptable ID with a photo and an expiry date.

- Undergo a palm vein scan for NCEES® and on-site identification. Entry and exit to the exam room require identification via the palm scan.

4) Prepare Food and Drink: You are allowed to store food and drink in a personal locker that is provided on-site. It is a good idea to bring caloric-dense foods that are also effective in boosting brain functionality, such as berries, dark chocolate and cocoa snacks, nuts, avocado toast, or eggs. Another way to do this, instead of taking five or more separate items, is to make a smoothie. Put 1 cup of rolled oats in a blender. Add a cup of yogurt, a banana, a handful of sunflower and/or pumpkin seeds, a handful of coconut flakes, a handful of dried raisins and sultanas, and a handful of mixed nuts, such as walnut, cashew, macadamia, and almonds, and blend it. You can add cream or milk if it's too thick. If you're up to it, add an egg or two. It's best to do this the night before, as the oats will absorb moisture and your smoothie may be too thick. If this is the case, add more yogurt, milk, or water. If you want a bit of texture variety, take a slab of dark chocolate and an energy snack bar separately.

5) Hydration: To increase mental performance, a neat trick is to bring a bottle of water from home with a sprinkle of salt in it.

***Note**: Always refer to the NCEES® website for the latest updates on exam procedures and what you can and cannot bring.

5.2 Time Management During The Exam

It is vitally important to understand the NCEES® FE Reference Handbook. When you are unsure, which will be rare given your preparation, the Reference Handbook will provide you with your answer. But to get there, you need to establish the key concept being covered in the question and search for the formula in the Reference Handbook.

Use the built in "flag" function on questions that are difficult to solve and come back to them later. Don't struggle with them because that will only frustrate you, make you anxious, and flustered. The idea is to complete what you know, and return later to complete the more difficult questions. If, after a minute of sitting on one problem and still not being sure how to approach it, chances are that you'll take longer than three minutes to solve it. Doing the problems you find easy or quick gives you a sense of empowerment, and you want to build on that in the first few hours.

Approach each problem with a process-of-elimination mindset. For the most part, each question will give you four possible answers. Start by looking for answers that just look/feel wrong, and ignore those. If you can narrow it down because of concepts that do not make sense, then do so. Are any of the answers similar to each other? Keep in mind that some of the problems will want you to select more than one option, so don't worry if more than one answer looks correct.

Avoid Overanalysis of exam questions. The questions are supposed to be answered in less than three minutes. They may look complex, but the problem is generally going to be clearly stated. Just read it slowly the first time. Do not waste time trying to analyze it for hidden meaning. If it doesn't make sense, flag it for later.

With 30 minutes left, do you review now, or solve the final 12 problems? You will know from the mock exams, just how much time you need to do a reasonable review. If you still have unsolved problems by the time you get to this point, decide whether it will be best to do the review or to get on with the unanswered problems. You may find the remaining problems are all in your areas of strength and can be done in 20 minutes. You will have an idea of how many questions you left unanswered, and you still have a few minutes to review. Most of all, trust your judgment because you know what you can do.

If you feel anxious or hit a mental block, take a break and do a breathing technique to help you regain focus. As soon as you find your mind wandering, or you're struggling to focus, pause. This happens in all tests, so embrace it as a short break. Sit up straight, hands in your lap, shoulders relaxed, and breathe in deeply for a count of five seconds. Hold your breath for five seconds. Breathe out for five seconds. Repeat this five or six times. Yes, it will take a minute and a half, but it slows your heart rate, oxygenates your blood, and gives you a fresher outlook. You will comfortably make up that 1.5 minute break through greater efficiency.

Slow it down. Yes, it sounds counter-intuitive for a tightly timed exam, but it helps you avoid silly errors. Read the problem statement slowly and carefully so you know exactly what they are asking you to solve. Write it down to be sure you understand it. You cannot solve the problem unless you understand what you are solving for.

As a measure of last resort, when you have eliminated what you can and are still left with more than one possible answer, take a guess. Not a wild guess because you just want to get to the next problem, but an educated guess. You've wrestled with this for three minutes already, and there is no penalty for wrong answers.

CHAPTER 6

Mathematics and Statistics

Mathematics and statistics form the foundational bedrock of electrical and computer engineering. A solid grasp of these subjects is crucial for understanding and solving a wide range of engineering problems. In the FE Electrical and Computer exam, candidates are expected to demonstrate proficiency in various mathematical and statistical concepts, including algebra, trigonometry, calculus, differential equations, linear algebra, and probability and statistics.

This chapter aims to equip you with the essential mathematical tools and statistical methods required for the exam. Each section is meticulously designed to break down complex topics into manageable and comprehensible chunks. We will start with the basics of algebra and trigonometry, exploring fundamental identities and laws that are pivotal in engineering calculations. We will then delve into the realm of complex numbers and vector analysis, crucial for understanding electrical circuits and signal processing.

6.1 Algebra and Trigonometry

Mathematics is the language of engineering, and a strong foundation in algebra and trigonometry is crucial for tackling a wide range of problems in electrical and computer engineering. This section will cover key concepts and identities that you need to understand and apply in your studies and on the FE exam.

Logarithmic Identities

Logarithms are used to simplify complex calculations involving exponential growth or decay, which are common in engineering. Understanding logarithmic identities can help you solve equations more efficiently.

- **Definition:** The logarithm of a number is the power to which a given base must be raised to produce that number. For example, in $log_b(a) = c, b^c = a$.
- **Common Bases:** The most common logarithms are base 10 (common logarithm) and base e (natural logarithm).

Key Identities

- **Product Rule:** $log_b(xy) = log_b(x) + log_b(y)$
- **Quotient Rule:** $log_b(\frac{x}{y}) = log_b(x) - log_b(y)$
- **Power Rule:** $log_b(x^y) = ylog_b(x)$
- **Change of Base Formula:** $log_b(x) = \frac{log_k(x)}{log_k(b)}$ where k is a new base, often 10 or e.

These identities simplify logarithmic expressions and are particularly useful when dealing with exponential functions in circuit analysis and signal processing.

Trigonometric Functions/Identities

Trigonometry deals with the relationships between the angles and sides of triangles, which is essential for analyzing waveforms, electrical circuits, and more.

Basic Functions

- **Sine ($\sin\theta$):** Opposite/Hypotenuse
- **Cosine ($\cos\theta$):** Adjacent/Hypotenuse
- **Tangent ($\tan\theta$):** Opposite/Adjacent

Key Identities

- **Pythagorean Identity:** $\sin^2\theta + \cos^2\theta = 1$
- **Angle Sum and Difference:**
  - $\sin(a \pm b) = \sin a \cos b \pm \cos a \sin b$
  - $\cos(a \pm b) = \cos a \cos b \mp \sin a \sin b$

Double Angle Formulas

  - $\sin(2\theta) = 2\sin\theta\cos\theta$
  - $\cos(2\theta) = \cos^2\theta - \sin^2\theta$

These identities help simplify complex trigonometric expressions, making it easier to solve engineering problems involving periodic functions, such as alternating current (AC) analysis.

Law of Sines

The Law of Sines is used to find unknown angles and sides of a triangle, especially when dealing with non-right triangles. This is particularly useful in engineering for resolving forces and analyzing vector quantities.

- **Formula:** $\frac{a}{\sin A} = \frac{b}{\sin B} = \frac{c}{\sin C}$
  - Here, a, b, and c are the lengths of the sides of a triangle, and A, B, and C are the opposite angles.

Application

- To find an unknown side: If you know two angles and one side, you can find the unknown side.

- To find an unknown angle: If you know two sides and one angle, you can find the unknown angle.

For example, in electrical engineering, you might use the Law of Sines to determine the resultant vector of multiple currents or forces acting at different angles.

Law of Cosines

The Law of Cosines is another crucial tool for solving triangles, particularly when the triangle is not a right triangle. It is especially useful in scenarios where the Law of Sines cannot be applied, such as when you know two sides and the included angle or all three sides of a triangle.

- **Formula:** $c^2 = a^2 + b^2 - 2abcosC$

    - Here, a, b, and c are the lengths of the sides of a triangle, and C is the included angle.

Application

- To find an unknown side: When you know two sides and the included angle, you can find the third side.

- To find an unknown angle: When you know all three sides, you can find any angle.

In practical electrical engineering problems, the Law of Cosines is often used in phasor analysis for AC circuits. For instance, it can help in calculating the magnitude and phase angle of the resultant current when multiple AC currents are combined.

6.2 Complex Numbers

In electrical and computer engineering, complex numbers play a crucial role in analyzing AC circuits, control systems, and signal processing. They help represent and solve equations involving two-dimensional quantities, such as voltage and current in AC circuits. This section will guide you through the basics of complex numbers and vector analysis, focusing on rectangular and polar forms and their operations.

Vector Analysis

Vector analysis with complex numbers is essential for understanding electrical phenomena that involve direction and magnitude. Complex numbers can be represented in two forms: rectangular and polar. Each form has its advantages, and understanding both is critical for solving various engineering problems.

Rectangular Form

The rectangular form of a complex number is written as a+bj, where a is the real part and bj is the imaginary part. The imaginary unit j (or i in mathematics) is defined as $\sqrt{-1}$.

- Example: 3+4j

Here, 3 is the real part, and 4j is the imaginary part.

To perform basic operations with complex numbers in rectangular form, you simply handle the real and imaginary parts separately.

Addition and Subtraction

- **Addition**

$$(a + bj) + (c + dj) = (a + c) + (b + d)j$$

Example

$$(3 + 4j) + (1 + 2j) = (3 + 1) + (4 + 2)j = 4 + 6j$$

- **Subtraction**

$$(a+bj)-(c+dj)=(a-c)+(b-d)j$$

Example

$$(3+4j)-(1+2j) = (3-1)+(4-2)j = 2+2j$$

Multiplication

- Multiply complex numbers as you would polynomials, applying the distributive property and using the fact that $j^2 = -1$.

Example

$$(3+4j)(1+2j) = 3\cdot 1 + 3\cdot 2j + 4j\cdot 1 + 4j\cdot 2j = 3+6j+4j+8j^2$$

Since

$$j^2 = -1, \textit{this becomes } 3+10j+8(-1) = 3+10j-8 = -5+10j$$

Division

- Division involves multiplying the numerator and denominator by the conjugate of the denominator.

Example

$$\frac{3+4j}{1+2j}$$

The conjugate of $1+2j$ *is* $1-2j$

Multiply the numerator and denominator by $1-2j$:

$$\frac{(3+4j)(1-2j)}{(1+2j)(1-2j)}$$

Simplify the denominator:

$$(1+2j)(1-2j) = 1-(2j)^2 = 1-4(-1) = 1+4 = 5$$

Expand the numerator:

$$(3+4j)(1-2j) = 3\cdot1 + 3\cdot(-2j) + 4j\cdot1 + 4j\cdot(-2j)$$

$$=3-6j+4j-8j^2$$

Since $j^2 = -1$

$= 3 - 6j + 4j + 8 = 3 + 8 - 6j + 4j = 11 - 2j$

Combine the results:

$$\frac{3 + 4j}{1 + 2j} = \frac{11 - 2j}{5} = \frac{11}{5} - \frac{2j}{5}$$

Final simplified form:

$$\frac{3 + 4j}{1 + 2j} = 2.2 - 0.4j$$

Polar Form and Operations

The polar form of a complex number represents it in terms of its magnitude and angle. It is written as $r\angle\theta$, where r is the magnitude and θ is the angle (in degrees or radians) from the positive real axis.

Example

$5\angle 53.13^\circ$

- Here, 5 is the magnitude and 53.13°is the angle.

To convert between rectangular and polar forms, use the following relationships:

- Magnitude: $r = \sqrt{(a^2 + b^2)}$
- Angle: $\theta = tan^{-1}(\frac{b}{a})$
- From polar to rectangular: $a = rcos\theta, b = rsin\theta$

Example Conversion

- Rectangular to Polar: For $3 + 4j$,
  - Magnitude: $r = \sqrt{3^2 + 4^2} = 9 + 16 = 5$
  - Angle: $\theta = tan^{-1}(\frac{4}{3}) = 53.13^\circ$
  - Polar Form: $5\angle 53.13^\circ$
- Polar to Rectangular: For $5\angle 53.13^\circ$,

- Real Part: $a = 5cos53.13^{\circ} = 3$
- Imaginary Part: $b = 5sin53.13^{\circ} = 4$
- Rectangular Form: $3 + 4j$

Operations in Polar Form

Multiplication and Division

- Multiplication: Multiply the magnitudes and add the angles.
  - $r1\angle\theta1 \times r2\angle\theta2 = (r1 \times r2)\angle(\theta1 + \theta2)$
  - **Example:** $5\angle30^{\circ} \times 2\angle45^{\circ} = 10\angle75^{\circ}$
- Division: Divide the magnitudes and subtract the angles.
  - $\frac{r1\angle\theta1}{r2\angle\theta2} = (\frac{r1}{r2})\angle(\theta1 - \theta2)$
  - **Example:** $\frac{5\angle60^{\circ}}{2\angle30^{\circ}} = 2.5\angle30^{\circ}$

Addition and Subtraction

- Addition and subtraction in polar form are not as straightforward. Convert the numbers to rectangular form, perform the operation, and convert back to polar form if necessary.

6.3 Calculus

Calculus is a branch of mathematics that deals with rates of change and the accumulation of quantities. It is essential for understanding and solving many engineering problems, particularly those involving dynamic systems. This section covers the basics of differential and integral calculus, providing you with the tools needed for the FE exam and your future engineering career.

Differential Calculus

Differential calculus focuses on the concept of the derivative, which represents the rate of change of a function. Understanding derivatives is crucial for analyzing and predicting the behavior of various engineering systems.

Definitions of the Derivative

The derivative of a function measures how the function's output changes as its input changes. In mathematical terms, the derivative of a function f(x) at a point x is defined as the limit: $f'(x) = \lim_{h \to 0} \frac{f(x+h) - f(x)}{h}$

This definition captures the idea of the instantaneous rate of change of the function at a given point. It's like finding the slope of the tangent line to the function at any specific point.

- **Example:** If $f(x) = x^2$, the derivative is:

$$f'(x) = \lim_{h \to 0} \frac{(x+h)^2 - x^2}{h} = \lim_{h \to 0} \frac{x^2 + 2xh + h^2 - x^2}{h} = \lim_{h \to 0} \frac{2xh + h^2}{h} = \lim_{h \to 0}(2x + h) = 2x$$

First and Second Derivatives

The first derivative of a function gives the rate of change of the function with respect to its variable. It can be interpreted as the slope of the tangent line to the function at any point. This is fundamental in understanding how quantities change and in predicting future behavior based on current trends.

- **Application:** In physics, the first derivative of the position function with respect to time gives the velocity of an object.

The second derivative of a function is the derivative of the first derivative. It measures the rate of change of the rate of change, providing information about the concavity of the function. The second derivative tells us how the rate of change itself is changing.

- **Example:** For $f(x) = x^3$,

  $f'(x) = 3x^2$

  $f''(x) = 6x$

- **Application:** In mechanics, the second derivative of the position function with respect to time gives the acceleration of an object. This is crucial for understanding forces and motion.

Understanding both the first and second derivatives allows engineers to analyze the stability and responsiveness of systems, optimize designs, and predict future states of dynamic systems.

Integral Calculus

Integral calculus is concerned with the accumulation of quantities and the areas under curves. It is the reverse process of differentiation and is essential for solving problems involving total quantities.

Indefinite Integrals

An indefinite integral, also known as an antiderivative, represents a family of functions whose derivative is the given function. The general form of an indefinite integral is: $\int f(x)\,dx$

- **Example:** The indefinite integral of 2x is: $\int 2x\,dx = x^2 + C$

  Here, C is the constant of integration, representing the family of all antiderivatives.

Basic Rules

- **Power Rule:** $\int x^n\,dx = \frac{x^{n+1}}{n+1} + C$
- **Sum Rule:** $\int (f(x) + g(x))\,dx = \int f(x)\,dx + \int g(x)\,dx$
- **Constant Multiple Rule:** $\int k \cdot f(x)\,dx = k\int f(x)\,dx$

These rules help integrate various functions, allowing you to find the original function given its derivative. This is particularly useful in physics and engineering for finding displacement from velocity or voltage from current.

Definite Integrals

A definite integral represents the accumulation of quantities over a specific interval. It calculates the net area under the curve of a function between two points. The general form of a definite integral is: $\int_a^b f(x)\,dx$

- **Example:** To find the area under the curve $f(x) = x^2$ $from\ x = 1\ to\ x = 3$:

$$\int_1^3 x^2\,dx = \left[\frac{x^3}{3}\right]_1^3 = \frac{3^3}{3} - \frac{1^3}{3} = \frac{27}{3} - \frac{1}{3} = 9 - \frac{1}{3} = \frac{26}{3}$$

The Fundamental Theorem of Calculus connects differentiation and integration, stating that if F(x) is the antiderivative of $f(x)$, then:

$$\int_a^b f(x)\,dx = F(b) - F(a)$$

This theorem simplifies the computation of definite integrals by using antiderivatives. Definite integrals have numerous applications in engineering:

- **Area Calculation:** Definite integrals can be used to find the area under a curve.
- **Physical Quantities:** They are used to calculate quantities such as displacement, work, and total charge.
- **Accumulated Change:** In control systems, definite integrals help determine the total accumulated change over time, such as the total energy consumed by a system.

Key Concepts and Applications

Differential Calculus

- Derivatives measure rates of change.
- The first derivative gives velocity; the second derivative gives acceleration.
- Important for motion, growth rates, and optimizing functions.

Integral Calculus

- Integrals accumulate quantities.
- Indefinite integrals provide antiderivatives.
- Definite integrals calculate areas and physical quantities like displacement and work.

6.4 Differential Equations

Differential equations relate a function to its derivatives. They are essential in engineering because they describe various physical phenomena, such as heat transfer, electrical circuits, and mechanical systems. This section will cover the basics of first- and second-order differential equations and distinguish between homogeneous and non-homogeneous solutions.

1st Order Differential Equations

First-order differential equations involve the first derivative of the function. They are often used to model simple systems and processes in engineering.

Definition and Examples

A first-order differential equation can be written in the form:

$$\frac{dy}{dx} + P(x)y = Q(x)$$

- **Example:** $\frac{dy}{dx} + 2y = 4$

This is a linear first-order differential equation.

Solving 1st Order Differential Equations

To solve a first-order differential equation, you can use several methods, one of which is the method of integrating factors. The integrating factor μ(x) is found by:

$\mu(x) = e^{\int P(x)dx}$

Multiplying the entire differential equation by the integrating factor transforms it into an exact differential equation that can be integrated directly.

- **Example**

  Given $\frac{dy}{dx} + 2y = 4$

  The integrating factor is $\mu(x) = e^{\int 2dx} = e^{2x}$

  Multiply through by e^{2x}: $\frac{e^{2x}dy}{dx} + 2e^{2x}y = 4e^{2x}$

  The left side becomes $\frac{d}{dx}(e^{2x}y) = 4e^{2x}$

  Integrate both sides: $e^{2x}y = 2e^{2x} + C$

  Solve for y: $y = 2 + Ce^{-2x}$

2nd Order Differential Equations

Second-order differential equations involve the second derivative of the function. These are often used in engineering to model more complex systems such as mechanical vibrations and electrical circuits.

Definition and Examples

A second-order differential equation can be written in the form:

$$\frac{d^2y}{dx^2} + P(x)\frac{dy}{dx} + Q(x)y = R(x)$$

- **Example:** $\frac{d^2y}{dx^2} - 3\frac{dy}{dx} + 2y = 0$

  This is a linear second-order homogeneous differential equation.

Solving 2nd Order Differential Equations

To solve a second-order differential equation, you often use characteristic equations if the coefficients are constant.

- **Example**

  Given $\frac{d^2y}{dx^2} - 3\frac{dy}{dx} + 2y = 0$

  Assume a solution of the form $y = e^{rx}$

  Substitute into the equation: $r^2e^{rx} - 3re^{rx} + 2e^{rx} = 0$

  Factor out e^{rx}: $e^{rx}(r^2 - 3r + 2) = 0$

  Solve the characteristic equation: $r^2 - 3r + 2 = 0$

  The roots are $r = 1 and\, r = 2r$

  The general solution is $y = C_1e^x + C_2e^{2x}$

Homogeneous and Non-Homogeneous Solutions

Differential equations can be classified as homogeneous or non-homogeneous depending on whether the right-hand side of the equation is zero.

Homogeneous Differential Equations

A homogeneous differential equation has the form:

$$\frac{d^2y}{dx^2} + P(x)\frac{dy}{dx} + Q(x)y = 0$$

The general solution to a homogeneous differential equation is the sum of the solutions of the characteristic equation.

- **Example:** $\frac{d^2y}{dx^2} + 4y = 0$

  The characteristic equation is $r2 + 4 = 0$

  The roots are $r = \pm 2i$

  The general solution is $y = C_1cos(2x) + C_2sin(2x)y$

Non-Homogeneous Differential Equations

A non-homogeneous differential equation has the form:

$$\frac{d^2y}{dx^2} + P(x)\frac{dy}{dx} + Q(x)y = R(x)$$

The general solution to a non-homogeneous differential equation is the sum of the complementary function (solution to the associated homogeneous equation) and a particular solution.

- **Example:** $\frac{d^2y}{dx^2} - 3\frac{dy}{dx} + 2y = e^{3x}$

  Solve the homogeneous part: $\frac{d^2y}{dx^2} - 3\frac{dy}{dx} + 2y = 0$

  The general solution to the homogeneous part is $y_h = C_1e^x + C_2e^{2x}$

  Find a particular solution y_p

  Assume $y_p = Ae^{3x}$

  Substitute y_p into the non-homogeneous equation to find A: $Ae^{3x}(9 - 9 + 2) = e^{3x} = A = \frac{1}{2}$

  Combine to get the general solution: $y = y_h + y_p = C_1e^x + C_2e^{2x} + \frac{1}{2}e^{3x}$

6.5 Linear Algebra

Linear algebra is a branch of mathematics that deals with vectors, vector spaces, and linear transformations. It is fundamental in many areas of engineering, including circuit analysis, control systems, and computer graphics. This section will cover linear equations, matrix operations, and an introduction to Laplace transforms.

Linear Equations

Linear equations represent straight lines. They are the simplest and most basic types of equations used in engineering.

Solving Linear Equations

A linear equation in one variable is typically written as ax+b=0 where aaa and b are constants. To solve for x, you simply isolate x on one side of the equation:

- **Example:** Solve $2x + 3 = 7$

  Subtract 3 from both sides: $2x = 4$

  Divide by 2: $x = 2$

In systems of linear equations, where you have multiple equations, the goal is to find values for the variables that satisfy all equations simultaneously. For example, in two variables, you might have:

$$\begin{cases} 2x + 3y = 5 \\ 4x - y = 1 \end{cases}$$

To solve this system, you can use methods such as substitution, elimination, or matrix operations.

Matrix Operations

Matrices are rectangular arrays of numbers that represent linear transformations. They are used to solve systems of linear equations and perform other linear algebra operations.

Basic Matrix Operations

- **Addition and Subtraction**
  - Matrices of the same dimension can be added or subtracted element-wise.
  - **Example:**

$$\begin{bmatrix} 1 & 2 \\ 3 & 4 \end{bmatrix} + \begin{bmatrix} 5 & 6 \\ 7 & 8 \end{bmatrix} = \begin{bmatrix} 6 & 8 \\ 10 & 12 \end{bmatrix}$$

- **Multiplication**
  - The product of two matrices A and B is defined if the number of columns in A equals the number of rows in B. The element in the i-th row and j-th column of the product is the dot product of the i-th row of A and the j-th column of B.
  - **Example:**

$$\begin{bmatrix} 1 & 2 \\ 3 & 4 \end{bmatrix} \times \begin{bmatrix} 2 & 0 \\ 1 & 2 \end{bmatrix} = \begin{bmatrix} (1 \cdot 2 + 2 \cdot 1) & 1 \cdot 0 + 2 \cdot 2) \\ (3 \cdot 2 + 4 \cdot 1) & (3 \cdot 0 + 4 \cdot 2) \end{bmatrix} = \begin{bmatrix} 4 & 4 \\ 10 & 8 \end{bmatrix}$$

- **Determinant**
  - The determinant is a scalar value that can be computed from a square matrix. It provides important properties about the matrix, such as whether it is invertible.
  - **Example for a 2x2 matrix**

$$\det \begin{bmatrix} a & b \\ c & d \end{bmatrix} = ad - bc$$

    - If the determinant is zero, the matrix is not invertible.

- **Inverse**
  - The inverse of a matrix A, denoted A^{-1}, is the matrix such that $AA^{-1} = I$, where I is the identity matrix. Not all matrices have inverses; a matrix must be square and have a non-zero determinant.
  - **Example for a 2x2 matrix**

$$A \begin{bmatrix} a & b \\ c & d \end{bmatrix} \implies A^{-1} = \frac{1}{ad - bc} \begin{bmatrix} d & -b \\ -c & a \end{bmatrix}$$

Laplace Transforms

Laplace transforms are a powerful tool used to transform differential equations into algebraic equations, which are easier to solve. This is particularly useful in control systems and signal processing.

Definition and Basic Properties

The Laplace transform of a function f(t) is defined as:

$$L\{f(t)\} = F(s) = \int_0^{\infty} e^{-st} f(t)\, dt$$

- **Linearity:** $L\{af(t) + bg(t)\} = aF(s) + bG(s)$
- **First Derivative:** $L\{f'(t)\} = sF(s) - f(0)$
- **Second Derivative:** $L\{f''(t)\} = s^2F(s) - sf(0) - f'(0)$

Applications

Laplace transforms simplify the process of solving linear differential equations by converting them into algebraic equations.

- **Example:** Solve $y'' + 3y' + 2y = 0$ with initial conditions $y(0) = 1 \ and \ y'(0) = 0.$
  - Take the Laplace transform of both sides: $L\{y'' + 3y' + 2y\} = L\{0\}$
  - Using the properties of Laplace transforms:

$$s^2Y(s) - sy(0) - y'(0) + 3(sY(s) - y(0)) + 2Y(s) = 0$$

$$s^2Y(s) - s \cdot 1 - 0 + 3(sY(s) - 1) + 2Y(s) = 0$$

$$s^2Y(s) + 3sY(s) + 2Y(s) - s - 3 = 0$$

$$Y(s)(s^2 + 3s + 2) = s + 3$$

$$Y(s) = \frac{s+3}{(s+1)\,(s+2)}$$

  - Perform partial fraction decomposition: $Y(s) = \frac{A}{s+1} + \frac{B}{s+2}$
  - Solve for A and B: $s + 3 = A(s + 2) + B(s + 1)$
  - After finding A and B, take the inverse Laplace transform to find y(t).

6.6 Statistics and Probability

Measures of Central Tendencies

Measures of central tendencies summarize a set of data by identifying the central point. The three main measures are mean, median, and mode.

- **Mean (Average):** Sum of all data points divided by the number of points.
  - **Formula:** $Mean = \sum \frac{x_i}{n}$
  - **Example:** For data points [2, 4, 6, 8, 10], the mean is $\frac{2+4+6+8+10}{5} = 6$.
- **Median:** The middle value when data points are arranged in ascending order.
  - **Calculation:** If the number of data points is odd, the median is the middle value. If even, it is the average of the two middle values.
  - **Example:** For [2, 4, 6, 8, 10], the median is 6. For [2, 4, 6, 8], the median is

$$\frac{4+6}{2} = 5$$

- **Mode:** The most frequently occurring value in a data set.
  - **Example:** For [2, 4, 4, 6, 8, 8, 8, 10], the mode is 8.

Measures of Dispersion

Measures of dispersion describe the spread of data points around the central tendency. Common measures include range, variance, and standard deviation.

- **Range:** The difference between the maximum and minimum values.
  - **Formula:** Range=Max−Min
  - **Example:** For $[2, 4, 6, 8, 10]$, the range is $10 - 2 = 8$.

- **Variance:** The average of the squared differences from the mean.
  - **Formula:** $Variance\ (Population): (\sigma^2) = \frac{\sum(x_i-\mu)^2}{n}$
  - **Example:** For [2, 4, 6, 8, 10], the variance is calculated as follows:
    1. Mean = 6
    2. Squared differences: $(2-6)^2, (4-6)^2, (6-6)^2, (8-6)^2, (10-6)^2$
    3. Variance = $\frac{16+4+0+4+16}{5} = 8$
- **Standard Deviation:** The square root of the variance, providing a measure of dispersion in the same units as the data.
  - **Formula:** $\sigma = \sqrt{\frac{\sum(x_i-\mu)^2}{N}}$
  - **Example:** For the variance of 8, the standard deviation is $\sqrt{8} \approx 2.83$

Probability Distributions

Probability distributions describe how the probabilities of a random variable are distributed. Key types include discrete and continuous distributions.

Discrete Distributions

- **Binomial Distribution:** Describes the number of successes in a fixed number of trials, with each trial having the same probability of success.
  - **Formula:** $P(X = k) = \binom{n}{k} p^k (1-p)^{n-k}$
  - **Example:** If a coin is flipped 10 times, the probability of getting exactly 6 heads is calculated using the binomial distribution.
- **Poisson Distribution:** Describes the probability of a given number of events happening in a fixed interval of time or space.
  - **Formula:** $P(X = k) = \frac{\lambda^k e^{-\lambda}}{k!}$

- **Example:** The number of emails received per hour in an office can be modeled using the Poisson distribution.

Continuous Distributions

- **Normal Distribution:** A continuous probability distribution characterized by a symmetric, bell-shaped curve.
  - **Properties:** Defined by mean (μ) and standard deviation (σ).
  - **Example:** Heights of people, test scores, and measurement errors often follow a normal distribution.
- **Exponential Distribution:** Describes the time between events in a Poisson process.
  - **Formula:** $f(x;\lambda) = \lambda e^{-\lambda x}$
  - **Example:** The time between arrivals of customers at a store can be modeled using the exponential distribution.

Expected Value

The expected value (mean) of a random variable is the long-run average value it takes on after many repetitions of the experiment.

- **Definition:** The sum of all possible values each multiplied by its probability.
- **Formula for Discrete Variables:** $E(X) = \sum x_i P(x_i)$
- **Formula for Continuous Variables:** $E(X) = \int_{-\infty}^{\infty} x f(x)\, dx$

Example

- For a discrete random variable representing the roll of a fair six-sided die, the expected value is:

$$E(X) = \sum_{i=1}^{6} i \cdot \frac{1}{6} = \frac{1+2+3+4+5+6}{6} = 3.5$$

Example

- In an engineering context, if a component has a 10% chance of failing within a year, the expected number of failures in a batch of 100 components is:

$$E(X) = 100 \times 0.1 = 10$$

Applying Statistics and Probability in Engineering

Let's consider a practical example where these concepts are applied in an electrical engineering context: the analysis of the lifespan of LED bulbs.

Scenario: An engineer wants to analyze the lifespan of a new batch of LED bulbs to ensure they meet quality standards. The engineer collects data on the lifespan (in hours) of 50 LED bulbs from the batch.

Step 1: Calculate Measures of Central Tendencies and Dispersions

- **Data Sample (in hours):** [900, 950, 1000, 1050, 1100, 950, 1000, 1050, 1100, 1150, 1000, 1050, 1100, 1150, 1200, 1050, 1100, 1150, 1200, 1250, 1100, 1150, 1200, 1250, 1300, 1150, 1200, 1250, 1300, 1350, 1200, 1250, 1300, 1350, 1400, 1250, 1300, 1350, 1400, 1450, 1300, 1350, 1400, 1450, 1500, 1350, 1400, 1450, 1500, 1550]

- **Mean**
    - Sum of lifespans = 58750
    - Number of bulbs = 50
    - Mean = $\frac{58750}{50} = 1175\ hours$

- **Median**
    - Ordered data: [900, 950, 950, 1000, 1000, 1000, 1050, 1050, 1050, 1050, 1100, 1100, 1100, 1100, 1150, 1150, 1150, 1150, 1200, 1200, 1200, 1200,

1250, 1250, 1250, 1250, 1300, 1300, 1300, 1300, 1350, 1350, 1350, 1350, 1400, 1400, 1400, 1400, 1450, 1450, 1450, 1450, 1500, 1500, 1500, 1500, 1550, 1550]

  - Middle values = 1150 and 1200
  - Median = $\frac{1150 + 1200}{2} = 1175$

- **Mode**
  - Most frequent lifespan = 1050 hours
- **Range**
  - Range = 1550 − 900 = 650 hours
- **Population Variance**
  - Population Variance = $\frac{Sum\ of\ squared\ differences}{50} \frac{225000}{50} = 4500\ hours^2$
- **Standard Deviation**
  - Standard Deviation = $\sqrt{4500} \approx 67.08\ hours$

Step 2: Probability Distributions

Assuming the lifespans follow a normal distribution:

- Mean (μ) = 1175 hours
- Standard Deviation (σ) = 67.08 hours

Step 3: Calculate Expected Value

If the engineer wants to predict the average lifespan of a bulb that fails:

- **Failure rate (10%)**
  - Expected number of failures in 100 bulbs = 100 * 0.1 = 10
  - If each failure costs $5 to replace, expected cost = 10×5=$50

ADDITIONAL ONLINE RESOURCES

- **1500+ PRACTICE PROBLEMS [4 PRACTICE EXAMS INCLUDED]**

Enhance your exam preparation with over 1000+ practice problems, complete with step-by-step solutions. Two weeks after your initial download, you will receive a weekly full-length practice exam with step-by-step solutions for four weeks, totaling over 1500 practice problems.

SCAN HERE

Having trouble accessing the resources? Email us at **info@learnovapublishing.com** — we reply within 24 hours, 7 days a week

- **VIDEO RESOURCE RECOMMENDATIONS**

Enhance your learning with our recommended video resources for a deeper understanding of the concepts. When using these resources, please check the current exam specifications.

An abundance of high-quality video resources is available for preparing for this section. For a comprehensive overview, we recommend the following:

YouTube Playlist: *FE Electrical and Computer: Review (Effective from July 2020) by Amit Roy*
Search for this playlist on YouTube. It covers almost all the exam topics, providing detailed explanations and condensing the material effectively.

Online Review Course: For a more organized approach, consider the comprehensive, free, video-based online review course called *"Electrical FE Review"* by Zach Stone, available at electricalfereview.com

Algebra and Trigonometry	
Key Concepts	**Handbook Page References**
Slope of a line	
Equation of a line (Slope-intercept form, Point-slope form, Standard form)	
Quadratic Equation	
Quadratic Surface	
Logarithmic Identities and Change of Base	
Trigonometric Identities	
Law of Sines and Law of Cosines	

Complex Numbers	
Key Concepts	**Handbook Page References**
Rectangular Form	
Polar Form	
Conversions between Rectangular and Polar Forms	
Addition, Subtraction, Multiplication, and Division	
Euler's Identity	

Discrete Mathematics	
Key Concepts	**Handbook Page References**
Set Theory	
Mapping Functions	
Directed Graphs	

Analytic Geometry	
Key Concepts	**Handbook Page References**
Distance Between Two Points	
Mensuration of Areas and Volumes:	
- Parabolic Segments	
- Ellipses	
- Circular Segments	
- Parallelograms	
- Regular Polygons	
Conic Sections:	
- Standard forms of Equations for Parabolas, Ellipses, Hyperbolas, and Circles	

Calculus	
Key Concepts	**Handbook Page References**
Differential Calculus:	
- Definition and Basic Rules of Differentiation	
- First and Second Derivatives	
- Applications: Maxima, Minima, and Inflection Points	
Integral Calculus:	
- Indefinite Integrals	
- Fundamental Theorem of Integral Calculus	
- Definite Integrals	
L'Hopital's Rule	

Differential Equations	
Key Concepts	**Handbook Page References**
1st Order Differential Equations:	
- Separation of Variables	
- Integrating Factor Methods	
2nd Order Differential Equations:	
- Characteristic Equation	
- Homogeneous and Non-Homogeneous Solutions	

Linear Algebra	
Key Concepts	**Handbook Page References**
Linear Equations	
Matrix Operations:	
- Addition, Subtraction, Multiplication	
- Determinant	
- Transpose	
- Inverse	

Vector Analysis	
Key Concepts	**Handbook Page References**
Vector Addition, Subtraction and Dot & Cross Product	
Vector Identities	

Measures of Central Tendencies and Dispersions	
Key Concepts	**Handbook Page References**
Mean (Arithmetic Mean)	
Mode	
Median	
Geometric Mean	
Standard Deviation	
Range	
Sample Variance	

Probability Distributions	
Key Concepts	**Handbook Page References**
Binomial Distribution:	
- Definition and Key Properties	
- Expected Value and Variance	
Normal Distribution:	
- Standard normal distribution	
- Z-Scores	
T-Distribution:	
- Hypothesis Testing	
- Degrees of Freedom	
Chi-Squared (χ^2) Distribution:	
- Goodness-Of-Fit Test	

Expected Value	
Key Concepts	**Handbook Page References**
Expected Value (Mean):	
- Discrete and Continuous Random Variables	
Variance:	
- Discrete and Continuous Random Variables	

NOTES/EQUATIONS SUMMARY

DIAGNOSTIC TEST CHECKLIST

Checklist Questions	Yes/No
Have you read and understood all the material in this chapter?	
Are you familiar with all the key concepts and formulas highlighted in the tables for this chapter?	
Have you filled in the tables with the correct NCEES® FE Reference Handbook page numbers for each listed concept and formula?	
Did you download and complete all the designated practice problems for this section?	
Did you review and understand the step-by-step solutions to the practice problems?	
Could you solve the practice problems without excessive reference to the solutions and within a reasonable time frame?	
Are you confident in applying the concepts and formulas from this chapter?	
Have you noted any areas of difficulty or topics that require further review before proceeding?	
Do you know how to quickly find information in the NCEES® FE Reference Handbook?	
Have you reviewed any errors or misunderstandings in this section's material to ensure clarity and understanding?	

CHAPTER 7

Ethics and Professional Practice

Ethics and professional practice are fundamental aspects of engineering that go beyond technical knowledge and skills. As an engineer, you have a responsibility not only to your employer and clients but also to the public and the environment. This chapter explores the ethical principles and professional practices that guide engineers in their work, ensuring safety, integrity, and accountability.

Understanding and adhering to ethical standards is crucial for maintaining public trust and advancing the engineering profession. Engineers are often faced with complex decisions that have significant impacts on society and the environment. Therefore, a solid grounding in ethics helps navigate these challenges and make decisions that are not only technically sound but also morally responsible.

7.1 Professional and Ethical Responsibilities

As an engineer, you have a duty to uphold public safety, demonstrate competence, and maintain accountability in all your professional activities. These responsibilities are foundational to the trust the public places in the engineering profession.

- **Public Safety:** Always ensure that your work prioritizes the safety, health, and welfare of the public. This means adhering to safety standards, regulations, and best practices in your designs and projects.

- **Competence:** Only take on tasks that you are qualified to perform. Continuously update your skills and knowledge to stay current with advancements in engineering practices and technologies.

- **Accountability:** Take full responsibility for your work. If mistakes occur, address them promptly and ethically, learning from these experiences to prevent future issues.

7.2 Codes of Ethics

Codes of ethics are crucial guidelines that help engineers navigate their professional duties with integrity and responsibility. These codes provide a framework for ethical decision-making, ensuring that engineers act in a manner that protects the public and upholds the profession's standards.

NSPE Code of Ethics

The National Society of Professional Engineers (NSPE) Code of Ethics outlines the fundamental principles that engineers must follow to ensure ethical practice.

- **Public Safety:** Always prioritize the safety, health, and welfare of the public in your work.

- **Competence:** Perform services only in your areas of expertise and continually update your skills.

- **Truthfulness:** Be honest and transparent in all public statements and reports.

- **Loyalty:** Act as faithful agents for your clients and employers, placing their interests above your own.

- **Integrity:** Avoid deceptive acts and conduct yourself honorably and responsibly.

IEEE Code of Ethics

The Institute of Electrical and Electronics Engineers (IEEE) Code of Ethics provides specific guidelines for professionals in electrical and electronics engineering.

- **Honesty and Realism:** Be honest and realistic in stating claims based on available data.
- **Conflict of Interest:** Avoid conflicts of interest and disclose any that may arise.
- **Improvement:** Seek to improve your technical competence and undertake only those tasks for which you are qualified.
- **Fairness:** Treat all persons fairly and with respect, regardless of race, religion, gender, disability, age, or national origin.

Adherence to Codes of Ethics

Following these codes of ethics is essential for maintaining the integrity and trustworthiness of the engineering profession.

- **Accountability:** Engineers are held accountable for their actions, and adherence to these codes ensures that they make decisions that are ethical and in the public's best interest.
- **Professional Reputation:** Upholding these ethical standards enhances your reputation and credibility as a professional engineer.

7.3 NCEES Model Law and Model Rules

The NCEES Model Law provides a framework for the licensing and regulation of engineers. It is designed to protect the public by ensuring that engineers meet a consistent standard of competence and ethical behavior.

- **Purpose:** The Model Law sets the foundation for state licensing laws, ensuring uniformity and consistency in the engineering profession across different jurisdictions.

- **Scope:** It covers the essential requirements for becoming a licensed engineer, including education, experience, and examinations, as well as guidelines for ethical practice and professional conduct.

Licensure Requirements

To become a licensed professional engineer, certain criteria must be met, as outlined in the NCEES Model Law.

- **Education:** Engineers must graduate from an accredited engineering program. Accreditation ensures that the program meets the standards necessary to prepare students for professional practice.

- **Experience:** Engineers must gain practical experience under the supervision of a licensed professional engineer. This typically involves a minimum of four years of progressive engineering experience.

- **Examinations:** Engineers must pass the Fundamentals of Engineering (FE) exam and the Principles and Practice of Engineering (PE) exam. These exams test the knowledge and skills required for competent practice.

Professional Conduct

Ethical behavior and professional responsibilities are core components of the NCEES Model Law. Engineers are expected to adhere to high standards of integrity and accountability.

- **Ethical Behavior:** Engineers must act with honesty, integrity, and fairness in all professional dealings. They should avoid conflicts of interest and ensure their actions do not harm the public.
- **Professional Responsibilities:** Engineers must perform services only in areas of their competence, continue their professional development, and ensure their work adheres to all relevant codes and standards.

Understanding Model Rules

The NCEES Model Rules provide detailed guidelines for professional practice, complementing the Model Law by offering more specific instructions on implementing its provisions.

- **Guidelines:** The Model Rules include detailed guidelines on topics such as professional competency, continuing education, ethical obligations, and disciplinary procedures.
- **Implementation:** These rules help ensure that the principles outlined in the Model Law are applied consistently and effectively across different states.

7.4 Intellectual Property and Safety

Intellectual property (IP) refers to creations of the mind, such as inventions, literary and artistic works, designs, and symbols. Protecting intellectual property is crucial for fostering innovation and maintaining ethical standards in engineering.

Types of Intellectual Property

There are several types of intellectual property, each serving different purposes and providing various levels of protection.

- **Patents:** Protect new inventions and processes, granting the patent holder exclusive rights to use, make, and sell the invention for a specified period.
- **Trademarks:** Protect brand names, slogans, and logos used to identify goods and services, ensuring that consumers can distinguish between different products and services.
- **Copyrights:** Protect original works of authorship, such as books, music, and software, giving the creator exclusive rights to use and distribute their work.
- **Trade Secrets:** Protect confidential business information, such as formulas, practices, and designs, that provide a competitive edge.

Respecting Intellectual Property

Respecting intellectual property is an essential ethical responsibility for engineers. This involves recognizing and honoring the IP rights of others.

- **Avoiding Plagiarism:** Always give proper credit to the original creators of ideas, designs, or data. Use citations and acknowledgments as needed.
- **Respecting Proprietary Information:** Do not use or disclose proprietary information without permission. Ensure that you have the necessary licenses or agreements to use patented technologies or copyrighted materials.

Safety Standards

Adhering to safety standards is critical in engineering practice to protect public health and welfare. Safety standards provide guidelines and regulations to ensure that engineering projects are safe for use.

- **Importance:** Following safety standards helps prevent accidents, injuries, and fatalities. It ensures that products and systems operate reliably and safely, protecting both users and the environment.
    - **Examples:** In construction, adhering to building codes ensures structural integrity. In electronics, following electrical safety standards prevents electrical hazards.

7.5 Safety

Safety is a fundamental aspect of engineering that ensures the well-being of the public and the environment. Engineers must always prioritize safety in their projects to prevent accidents and protect lives. Ensuring safety in engineering projects helps prevent injuries, fatalities, and environmental damage. It also builds public trust and confidence in engineering solutions.

Adherence to Safety Regulations

Engineers must follow established safety codes and regulations in all their work. These guidelines provide a framework for designing and implementing safe engineering solutions.

- **Safety Codes:** Adhering to industry-specific codes, such as building codes and electrical standards, is crucial.

- **Regulations:** Compliance with local, national, and international regulations ensures that engineering projects meet all required safety standards.

Risk Management

Risk management involves identifying potential hazards and taking steps to mitigate them. This proactive approach helps ensure the safety and reliability of engineering projects.

- **Identifying Hazards:** Engineers must assess projects for potential risks, such as structural weaknesses or electrical faults.
- **Mitigation Strategies:** Implementing measures to reduce or eliminate risks, such as safety barriers, alarms, and regular maintenance.

ADDITIONAL ONLINE RESOURCES

- **1500+ PRACTICE PROBLEMS [4 PRACTICE EXAMS INCLUDED]**

Enhance your exam preparation with over 1000+ practice problems, complete with step-by-step solutions. Two weeks after your initial download, you will receive a weekly full-length practice exam with step-by-step solutions for four weeks, totaling over 1500 practice problems.

SCAN HERE

Having trouble accessing the resources? Email us at **info@learnovapublishing.com** — we reply within 24 hours, 7 days a week

- **VIDEO RESOURCE RECOMMENDATIONS**

Enhance your learning with our recommended video resources for a deeper understanding of the concepts. When using these resources, please check the current exam specifications.

An abundance of high-quality video resources is available for preparing for this section. For a comprehensive overview, we recommend the following:

YouTube Playlist: *FE Electrical and Computer: Review (Effective from July 2020) by Amit Roy*
Search for this playlist on YouTube. It covers almost all the exam topics, providing detailed explanations and condensing the material effectively.

Online Review Course: For a more organized approach, consider the comprehensive, free, video-based online review course called *"Electrical FE Review"* by Zach Stone, available at electricalfereview.com

Code of Ethics (Professional and Technical Societies)	
Key Concepts	**Handbook Page References**
Background and Definitions	
Rules of Professional Conduct	

NCEES Model Law and Model Rules	
Key Concepts	**Handbook Page References**
Licensure Requirements	
Licensee's Obligations	
Grounds for Disciplinary Action	
Certificate of Authorization Requirements	
Exemption Clause	

Intellectual Property and Safety	
Key Concepts	**Handbook Page References**
Intellectual Property:	
- Copyright	
- Patent	
- Trademark	
- Industrial Design	
- Trade Secret	
Safety:	
- Electrical Safety	
- Grounding	
- Material Safety Data Sheets (MSDS)	
- Personal Protective Equipment (PPE)	

NOTES/EQUATIONS SUMMARY

DIAGNOSTIC TEST CHECKLIST

Checklist Questions	Yes/No
Have you read and understood all the material in this chapter?	
Are you familiar with all the key concepts and formulas highlighted in the tables for this chapter?	
Have you filled in the tables with the correct NCEES® FE Reference Handbook page numbers for each listed concept and formula?	
Did you download and complete all the designated practice problems for this section?	
Did you review and understand the step-by-step solutions to the practice problems?	
Could you solve the practice problems without excessive reference to the solutions and within a reasonable time frame?	
Are you confident in applying the concepts and formulas from this chapter?	
Have you noted any areas of difficulty or topics that require further review before proceeding?	
Do you know how to quickly find information in the NCEES® FE Reference Handbook?	
Have you reviewed any errors or misunderstandings in this section's material to ensure clarity and understanding?	

CHAPTER 8

Engineering Economics

Engineering economics is the application of economic principles to engineering projects. It involves evaluating the costs and benefits of engineering decisions to ensure that resources are used efficiently and effectively. This chapter will provide you with the tools and techniques needed to analyze the economic aspects of engineering projects, helping you make informed decisions that balance technical and financial considerations.

8.1 Economic Analysis Methods

Economic analysis methods are crucial tools for engineers. They help evaluate the feasibility, efficiency, and overall value of engineering projects. By understanding and applying these methods, engineers can make informed decisions that balance technical and financial considerations, ensuring that resources are used effectively and projects are economically viable.

Economic analysis is essential for several reasons:

- **Project Feasibility:** Determines whether a project is worth pursuing based on its costs and potential benefits.
- **Resource Allocation:** Helps allocate resources efficiently to maximize project outcomes.
- **Decision Making:** Provides a systematic approach to compare different project options and choose the best one.

Cost Analysis

Cost analysis involves identifying and evaluating all costs associated with a project. This helps in budgeting and managing expenses effectively.

- **Fixed Costs:** These costs remain constant regardless of the level of production or project activity.
  - Examples: Rent, salaries, and equipment depreciation.
  - **Example:** A factory's monthly rent remains the same regardless of how many units it produces.
- **Variable Costs:** These costs vary directly with the level of production or project activity.
  - Examples: Raw materials and utility costs.
  - **Example:** The cost of raw materials increases as more products are manufactured.
- **Total Cost Calculation:** The sum of fixed and variable costs gives the total cost of the project.
  - **Formula:** Total Cost = Fixed Costs + Variable Costs

The Time Value of Money

The time value of money (TVM) is a fundamental concept in engineering economics. It recognizes that a dollar today is worth more than a dollar in the future due to its potential earning capacity.

- **Present Value (PV):** The current value of a future sum of money, discounted at a specific interest rate.
  - **Formula:** $PV = \frac{FV}{(1+r)^n}$
  - **Example:** If you will receive $1,000 in 5 years and the annual discount rate is 5%, the present value is: $PV = \frac{1000}{(1+0.05)^5} = \frac{1000}{1.276} \approx 784.96$
- **Future Value (FV):** The value of a current sum of money at a future date, compounded at a specific interest rate.
  - **Formula:** $FV = PV \times (1 + r)^n$
  - **Example:** If you invest $1,000 today at an annual interest rate of 5%, the future value in 5 years is: $FV = 1000 \times (1 + 0.05)^5 = 1000 \times 1.276 = 1276$

Economic Decision Making

Economic decision-making involves comparing the costs and benefits of different project options to determine the best course of action. Several techniques are used in this process:

- **Cost-Benefit Analysis (CBA):** A systematic approach to estimating the strengths and weaknesses of alternatives by comparing their costs and benefits.
  - **Steps**
    1. Identify all costs and benefits associated with each alternative.

2. Assign monetary values to costs and benefits.
3. Calculate the net benefit by subtracting total costs from total benefits.
4. Choose the alternative with the highest net benefit.

  - **Example:** Choosing between two types of machinery for a manufacturing plant. Machine A costs $10,000 and provides $15,000 in benefits, while Machine B costs $8,000 and provides $12,000 in benefits. Machine A has a higher net benefit ($5,000) compared to Machine B ($4,000).

- **Break-Even Analysis:** Determines the point at which total revenues equal total costs, indicating no profit or loss.
  - **Formula:** $Break - Even\ Point\ (units) = \frac{FixedCosts}{Selling\ Price\ per\ Unit - Variable\ Cost\ per\ Unit}$
  - **Example:** If fixed costs are $5,000, the selling price per unit is $50, and the variable cost per unit is $30, the break-even point is:

$$\frac{5000}{50-30} = \frac{5000}{20} = 250\ units$$

- **Payback Period:** The time it takes for an investment to generate enough cash flow to recover its initial cost.
  - **Formula:** $Payback\ Period\ = \frac{Initial\ Investment}{Annual\ Cash\ Inflow}$
  - **Example:** If an investment costs $10,000 and generates $2,500 annually, the payback period is: $\frac{10000}{2500} = 4\ years$

Depreciation Methods

Depreciation is the process of allocating the cost of a tangible asset over its useful life. This is important for understanding the true cost and value of assets over time.

- **Straight-Line Depreciation:** The simplest method, where an equal amount of depreciation is allocated each year over the asset's useful life.
  - **Formula:** $Depreciation\ Expense\ = \frac{Cost - Salvage\ Value}{Useful\ Life}$
  - **Example:** If an asset costs $10,000, has a salvage value of $2,000, and a useful life of 8 years, the annual depreciation is:

$$\frac{10000 - 2000}{8} = 1000$$

- **Declining Balance Depreciation:** An accelerated method where a higher amount of depreciation is recorded in the earlier years of the asset's life.
  - **Formula:** $Depreciation\ Expense\ =\ BookValue \times \frac{2}{Useful\ Life}$
  - **Example:** If an asset has a book value of $10,000 and a useful life of 5 years, the first year's depreciation using double declining balance is:

$$10000 \times \frac{2}{5} = 4000$$

- **Sum-of-the-Years-Digits (SYD):** Another accelerated method where depreciation is based on the sum of the years' digits.
  - **Formula:** $Depreciation\ Expense\ = \frac{Remaining\ Life}{SYD} \times (Cost\ -\ Salvage\ Value)$
  - **Example:** For an asset with a useful life of 5 years, the SYD is 1+2+3+4+5=15. If the asset costs $10,000 with a salvage value of $2,000, the first year's depreciation is: $\frac{5}{15} \times (10000 - 2000) = \frac{5}{15} \times 8000 = 2666.67$

8.2 Financial Decision-Making (Time Value of Money, Costs, Uncertainty)

Time Value of Money

The time value of money (TVM) is a key concept in financial decision-making. It recognizes that money available today is worth more than the same amount in the future due to its potential earning capacity. This principle is essential for evaluating investments, comparing financial options, and making informed economic decisions in engineering projects.

Explanation and Importance

- **Time Value:** Money today can be invested to earn interest or returns, making it more valuable than money received in the future.
- **Decision-Making:** TVM helps in assessing the value of future cash flows, comparing different investment opportunities, and determining the feasibility of projects.

Present Value (PV) Calculations

Present value is the current worth of a future sum of money, discounted at a specific interest rate. It helps in understanding how much future cash flows are worth today.

Formula: $PV = \frac{FV}{(1+r)^n}$

Example: If you will receive \$1,000 in 5 years and the annual discount rate is 5%, the present value is: $PV = \frac{1000}{(1+0.05)^5} = \frac{1000}{1.276} \approx 784.96$

Explanation: This means that \$784.96 today is equivalent to \$1,000 received in 5 years at a 5% discount rate.

Future Value (FV) Calculations

Future value is the amount of money an investment will grow to over a period of time at a given interest rate. It helps in understanding the potential growth of current investments.

Formula: $FV = PV \times (1 + r)^n$

Example: If you invest $1,000 today at an annual interest rate of 5%, the future value in 5 years is: $FV = 1000 \times (1 + 0.05)^5 = 1000 \times 1.276 = 1276$

Explanation: This shows that $1,000 invested today will grow to $1,276 in 5 years with a 5% annual interest rate.

Applications in Engineering Projects

TVM is widely used in engineering projects to evaluate the economic feasibility of investments, compare project alternatives, and make decisions about project financing and scheduling.

- **Investment Analysis:** Determining the value of future returns from investing in new technologies or infrastructure.
    - **Example:** Analyzing whether to invest in energy-efficient equipment that costs $50,000 now but is expected to save $70,000 in energy costs over 10 years. Using PV calculations, engineers can determine if the investment is worthwhile.
- **Project Evaluation:** Comparing the present value of project costs and benefits to decide whether to proceed with a project.
    - **Example:** Calculating the present value of expected revenues from a new product line to see if it exceeds the initial investment.
- **Cash Flow Management:** Planning and managing project cash flows to ensure financial stability and profitability.

- **Example:** Managing construction project payments to ensure positive cash flow throughout the project duration.

Costs

Understanding and managing costs is crucial for the success of engineering projects. This involves identifying different types of costs, estimating them accurately, and controlling them effectively to ensure projects stay within budget.

Types of Costs: Fixed, Variable, and Total

- **Fixed Costs:** Costs that remain constant regardless of the level of production or project activity.
  - **Examples:** Rent, salaries, insurance.
  - **Detailed Example:** A manufacturing plant has a fixed monthly rent of \$5,000, which does not change whether the plant produces 100 units or 1,000 units of product.
- **Variable Costs:** Costs that vary directly with the level of production or project activity.
  - **Examples:** Raw materials, utility costs.
  - **Detailed Example:** If each unit of product requires \$20 of raw materials, then producing 100 units costs \$2,000 and producing 1,000 units costs \$20,000 in raw materials.
- **Total Costs:** The sum of fixed and variable costs, representing the total expenditure for a project.
  - **Formula:** Total Cost = Fixed Costs + Variable Costs
  - **Example:** If the fixed costs are \$5,000 and the variable costs for producing 200 units are \$4,000, then the total cost is:

$$Total\ Cost = 5000 + 4000 = 9000$$

Cost Estimation Techniques

Accurate cost estimation is essential for budgeting, planning, and managing engineering projects. Various techniques can be used to estimate costs effectively.

- **Analogous Estimating:** Using historical data from similar projects to estimate costs.
  - **Detailed Example:** If a previous project of similar scope costs $100,000, a new project with similar requirements might be estimated to cost around the same amount.
- **Parametric Estimating:** Using statistical models to estimate costs based on project parameters.
  - **Detailed Example:** If building a road costs $500,000 per mile, then building a 10-mile road is estimated to cost $5,000,000.
- **Bottom-Up Estimating:** Estimating costs by breaking down the project into smaller components and estimating the cost of each component.
  - **Detailed Example:** Estimating the cost of materials, labor, and equipment separately for each phase of a construction project, then summing them up to get the total project cost.

Analyzing and Controlling Project Costs

Effective cost management involves not only estimating costs but also analyzing and controlling them throughout the project lifecycle.

- **Cost Analysis:** Regularly comparing actual costs with estimated costs to identify variances and take corrective actions.

- **Detailed Example:** If the estimated cost for materials was $10,000 but the actual cost is $12,000, identify the reasons for the increase and adjust the project plan or budget as needed.

- **Cost Control:** Implementing measures to keep costs within budget, such as cost-saving initiatives, efficient resource allocation, and waste reduction.
  - **Detailed Example:** Reducing material waste through better inventory management or negotiating better prices with suppliers to reduce costs.

Uncertainty

Uncertainty is inherent in engineering projects due to various factors such as changing market conditions, technological advancements, and unforeseen events. Managing uncertainty is crucial for making informed decisions and ensuring project success.

Dealing with Uncertainty in Engineering Economics

Uncertainty can affect project costs, schedules, and outcomes. Engineers need to identify potential uncertainties and develop strategies to manage them effectively.

- **Scenario Analysis:** Evaluating different scenarios to understand the potential impact of uncertainties on project outcomes.
  - **Detailed Example:** Analyzing best-case, worst-case, and most likely scenarios for project costs and timelines. For instance, in a construction project, considering scenarios where the cost of materials may increase due to market fluctuations.

- **Sensitivity Analysis:** Assessing how changes in key variables affect project outcomes.
  - **Detailed Example:** Understanding how variations in raw material prices or labor costs impact the total project cost. For example, if the price of

steel increases by 10%, how does that affect the overall budget of a construction project?

Risk Analysis and Management

Risk analysis involves identifying, assessing, and prioritizing risks, while risk management involves developing strategies to mitigate and manage these risks.

- **Risk Identification:** Identifying potential risks that could impact the project.
  - **Examples:** Supply chain disruptions, regulatory changes, and technical failures.
  - **Detailed Example:** Identifying the risk of delays in material delivery and its impact on the project schedule.
- **Risk Assessment:** Evaluating the likelihood and impact of identified risks.
  - **Detailed Example:** Assessing the probability of a supplier failing to deliver materials on time and its impact on the project timeline.
- **Risk Mitigation:** Developing strategies to reduce the likelihood or impact of risks.
  - **Detailed Example:** Diversifying suppliers to reduce dependency on a single supplier or implementing backup plans for critical activities.

Decision-Making Under Uncertainty

Decision-making under uncertainty involves evaluating options and making choices that maximize the likelihood of achieving project goals despite uncertainties.

- **Expected Value Analysis:** Calculating the expected value of different options based on their probabilities and outcomes.

- **Detailed Example:** Choosing between two investment options by calculating their expected returns and selecting the one with the higher expected value. For instance, if Option A has a 70% chance of earning $10,000 and a 30% chance of losing $2,000, the expected value is:

  $$EV = 0.7 \times 10000 + 0.3 \times (-2000) = 7000 - 600 = 6400$$

- **Decision Trees:** Using graphical representations to analyze decision options and their possible consequences.
  - **Detailed Example:** A decision tree for selecting a project alternative, considering factors such as costs, benefits, and risks. For example, evaluating whether to invest in a new technology by mapping out potential outcomes and their probabilities.

ADDITIONAL ONLINE RESOURCES

- **1500+ PRACTICE PROBLEMS [4 PRACTICE EXAMS INCLUDED]**

Enhance your exam preparation with over 1000+ practice problems, complete with step-by-step solutions. Two weeks after your initial download, you will receive a weekly full-length practice exam with step-by-step solutions for four weeks, totaling over 1500 practice problems.

SCAN HERE

Having trouble accessing the res[illegible]nfo@learnovapublishing.com — we reply within 24 hours, 7 days a week

- **VIDEO RESOURCE RECOMMENDATIONS**

Enhance your learning with our recommended video resources for a deeper understanding of the concepts. When using these resources, please check the current exam specifications.

An abundance of high-quality video resources is available for preparing for this section. For a comprehensive overview, we recommend the following:

YouTube Playlist: *FE Electrical and Computer: Review (Effective from July 2020) by Amit Roy*
Search for this playlist on YouTube. It covers almost all the exam topics, providing detailed explanations and condensing the material effectively.

Online Review Course: For a more organized approach, consider the comprehensive, free, video-based online review course called *"Electrical FE Review"* by Zach Stone, available at electricalfereview.com

Time Value of Money	
Key Concepts	**Handbook Page References**
Single Payment Equivalence:	
- Single Payment Future Worth	
- Single Payment Present Worth	
Uniform Series Equivalence:	
- Uniform Series Future Worth	
- Uniform Series Annual Value of a Sinking Fund	
- Uniform Series Present Worth	
- Uniform Series Annual Value Using the Capital Recovery Factor	
Uniform Gradient Equivalence:	
- Uniform Gradient Present Worth	
- Uniform Gradient Future Worth	
- Uniform Gradient Uniform Series Factor	
Nonannual Compounding:	
- Effective Annual Interest Rate	

Cost Estimation and Depreciation	
Key Concepts	**Handbook Page References**
Cost Estimation:	
- Inflation	
- Capitalized Costs	
Depreciation:	
- Straight Line Method	
- Modified Accelerated Cost Recovery System (MACRS)	
Book Value	

Economic Analyses	
Key Concepts	**Handbook Page References**
Break-Even Analysis	
Benefit-Cost Analysis	

Risk Identification	
Key Concepts	**Handbook Page References**
Definition	
Rate-of-Return	
Decision Tree	

NOTES/EQUATIONS SUMMARY

DIAGNOSTIC TEST CHECKLIST

Checklist Questions	Yes/No
Have you read and understood all the material in this chapter?	
Are you familiar with all the key concepts and formulas highlighted in the tables for this chapter?	
Have you filled in the tables with the correct NCEES® FE Reference Handbook page numbers for each listed concept and formula?	
Did you download and complete all the designated practice problems for this section?	
Did you review and understand the step-by-step solutions to the practice problems?	
Could you solve the practice problems without excessive reference to the solutions and within a reasonable time frame?	
Are you confident in applying the concepts and formulas from this chapter?	
Have you noted any areas of difficulty or topics that require further review before proceeding?	
Do you know how to quickly find information in the NCEES® FE Reference Handbook?	
Have you reviewed any errors or misunderstandings in this section's material to ensure clarity and understanding?	

CHAPTER 9

Electrical Circuits

Electrical circuits are fundamental to the field of electrical and computer engineering. Understanding how electrical circuits work is essential for designing and analyzing a wide range of systems, from simple household electronics to complex industrial machinery. This chapter will provide a comprehensive overview of both direct current (DC) and alternating current (AC) circuits, covering key concepts, principles, and analysis techniques.

9.1 DC and AC Circuits

DC Circuits

Direct current (DC) circuits are electrical circuits in which the current flows in one constant direction. They are fundamental in various applications, including batteries, electronics, and certain power systems. Understanding DC circuits involves grasping the basic principles of voltage, current, and resistance and how they interact within a circuit.

Ohm's Law

Ohm's Law is a fundamental principle in electrical engineering that relates voltage (V), current (I), and resistance (R) in a circuit. It is expressed as: $V = I \times R$

- **Voltage (V):** The potential difference between two points in a circuit.
- **Current (I):** The flow of electric charge through a conductor, measured in amperes (A).
- **Resistance (R):** The opposition to the flow of current, measured in ohms (Ω).

Example

- If a resistor with a resistance of 5 ohms (Ω) has a current of 2 amperes (A) flowing through it, the voltage across the resistor can be calculated as:

$$V = I \times R = 2 \times 5 = 10\ volts$$

Series and Parallel Circuits

DC circuits can be arranged in series or parallel configurations, each with distinct characteristics.

Series Circuits

- Components are connected end-to-end, so there is only one path for current flow.
- The total resistance (R_{total}) is the sum of individual resistances.

$$R_{total} = R1 + R2 + R3 + \ldots.$$

Example

- If three resistors with resistances of 2Ω, 3Ω, and 5Ω are connected in series, the total resistance is: $R_{total} = 2 + 3 + 5 = 10\ ohms$
- The same current flows through all components.

Parallel Circuits

- Components are connected across the same two points, providing multiple paths for current flow.
- The total resistance (R_{total}) is found using the reciprocal of the sum of the reciprocals of individual resistances.

$$\frac{1}{R_{total}} = \frac{1}{R1} + \frac{1}{R2} + \frac{1}{R3} + \cdots$$

Example

- If three resistors with resistances of 2Ω, 3Ω, and 6Ω are connected in parallel, the total resistance is:

$$\frac{1}{Rtotal} = \frac{1}{2} + \frac{1}{3} + \frac{1}{6} = \frac{3}{6} + \frac{2}{6} + \frac{1}{6} = 1$$

$$R_{total} = 1\ ohm$$

- The voltage across each component is the same, but the current divides among the parallel paths.

Kirchhoff's Laws

Kirchhoff's Laws are two essential principles for analyzing complex DC circuits: Kirchhoff's Voltage Law (KVL) and Kirchhoff's Current Law (KCL).

Kirchhoff's Voltage Law (KVL)

- States that the sum of all voltages around any closed loop in a circuit must equal zero: $\sum V = 0$

Example

- In a closed loop with a battery of 12V, a resistor of 4Ω, and another resistor of 2Ω, the voltage drops across the resistors must sum to 12V.

  $V1 + V2 = 12V$

  With current I through the resistors: $V1 = I \times 4\ and\ V2 = I \times 2$

Applying KVL: $4I + 2I = 12 \Rightarrow 6I = 12 \Rightarrow I = 2A$

Kirchhoff's Current Law (KCL)

- States that the sum of currents entering a junction must equal the sum of currents leaving the junction. $\sum I_{in} = \sum I_{out}$

Example

- At a junction where three branches meet, with currents of 3A, 2A, and 1A flowing into the junction, and one branch with an unknown current I flowing out: $3 + 2 + 1 = I \Rightarrow 6 = I$

The current flowing out of the junction is 6A.

Examples and Applications of DC Circuits

Example 1

- Consider a series circuit with a 9V battery, a 2Ω resistor, and a 3Ω resistor.
    - Total resistance: $R_{total} = 2 + 3 = 5\ ohms$
    - Total current: $I = \frac{V}{R} = \frac{9}{5} = 1.8\ A$
    - Voltage drops: $V_1 = 1.8 \times 2 = 3.6V\ and\ V_2 = 1.8 \times 3 = 5.4V$

Example 2

- Consider a parallel circuit with a 12V battery, a 4Ω resistor, and a 6Ω resistor.
    - Total resistance: $\frac{1}{R_{total}} = \frac{1}{4} + \frac{1}{6} = \frac{3}{12} + \frac{2}{12} = \frac{5}{12} \Rightarrow R_{total} = \frac{12}{5} = 2.4\ ohms$
    - Total current: $I_{total} = \frac{12}{2.4} = 5\ A$
    - Current through each resistor: $I_1 = \frac{12}{4} = 3\ A\ and\ I_2 = \frac{12}{6} = 2\ A$

Applications

- **Electronics:** DC circuits are used in batteries powering devices like remote controls, flashlights, and smartphones.

- **Automotive:** Car electrical systems use DC circuits for components such as lights, radios, and starters.
- **Renewable Energy:** Solar panels generate DC power, which is then converted to AC for home use.

AC Circuits

Introduction to AC Circuits

Alternating current (AC) circuits are a type of electrical circuit in which the current periodically reverses direction. Unlike direct current (DC) circuits, which have a constant flow of current, AC circuits are characterized by a flow of current that changes direction at regular intervals. AC circuits are used widely in household electrical systems, power distribution networks, and various electronic devices because they are efficient for long-distance transmission and can easily be transformed to different voltage levels.

Sinusoidal Waveforms

In AC circuits, the most common type of waveform is the sinusoidal waveform. A sinusoidal waveform is a smooth, repetitive oscillation that can be described by the sine function.

- **Characteristics:**
  - **Amplitude (A):** The maximum value of the waveform, either positive or negative.
  - **Frequency (f):** The number of cycles per second, measured in hertz (Hz).
  - **Period (T):** The time it takes to complete one cycle, which is the reciprocal of frequency ($T = 1/f$).
  - **Phase (φ):** The offset of the waveform relative to a reference point.

Example

- If a sinusoidal voltage has an amplitude of 120V and a frequency of 60Hz, its equation can be written as: $v(t) = 120sin(2\pi \times 60t)$
    - This means the voltage oscillates between +120V and -120V, completing 60 cycles every second.

RMS and Peak Values

In AC circuits, two important values are the root mean square (RMS) value and the peak value of a waveform.

- **Peak Value (V_{peak}):** The maximum value of the waveform.
    - **Example:** For a sinusoidal waveform with a peak value of 170V, the waveform oscillates between +170V and -170V.
- **RMS Value (V_{rms}):** The effective value of the AC voltage or current, equivalent to a DC value that delivers the same power. For a sinusoidal waveform, the RMS value is given by: $V_{rms} = \frac{V_{peak}}{\sqrt{2}}$
    - **Example:** If the peak value is 170V, the RMS value is:

$$V_{rms} = 170/\sqrt{2} \approx 120V$$

Impedance and Reactance

Impedance and reactance are key concepts in AC circuits that describe the opposition to the flow of current.

- **Impedance (Z):** The total opposition to current flow in an AC circuit, measured in ohms (Ω). It includes both resistance (R) and reactance (X).
    - **Formula:** $Z = \sqrt{R^2 + X^2}$

- **Reactance (X):** The opposition to current flow due to inductance (L) and capacitance (C).
  - **Inductive Reactance (X_L):** $X_L = 2\pi f L$
  - **Capacitive Reactance (X_C):** $X_C = \frac{1}{2\pi f C}$

Example

- If an AC circuit has a resistor of 10Ω and an inductor with an inductive reactance of 15Ω, the impedance is:

$$Z = \sqrt{10^2 + 15^2} = \sqrt{100 + 225} = 325 \approx 18.03\ \Omega$$

Power in AC Circuits

Power in AC circuits can be divided into three types: real power (P), reactive power (Q), and apparent power (S).

- **Real Power (P):** The actual power consumed by the circuit, measured in watts (W). It is given by: $P = V_{rms} I_{rms} cos(\phi)$
  - **Example:** If $V_{rms} = 120V, I_{rms} = 10A$, and the phase angle ϕ is 30 degrees, then: $P = 120 \times 10 \times cos(30^{\circ}) = 1200 \times 0.866 = 1039.2W$
- **Reactive Power (Q):** The power stored and released by inductors and capacitors, measured in volt-amperes reactive (VAR). It is given by:

$$Q = V_{rms} I_{rms} sin(\phi)$$

  - **Example:** Using the same values as above:

$$Q = 120 \times 10 \times sin(30^{\circ}) = 1200 \times 0.5 = 600\ VAR$$

- **Apparent Power (S):** The product of the RMS voltage and RMS current, measured in volt-amperes (VA). It represents the total power in the circuit, combining both real and reactive power. $S = V_{rms} I_{rms}$

Example:

For $V_{rms} = 120V\ and\ I_{rms} = 10A$:

$$S = 120 \times 10 = 1200\ VA$$

Examples and Applications of AC Circuits

Example 1

- Consider an AC circuit with a 120V RMS voltage source, a 10Ω resistor, and a 0.02H inductor in series.
  - **Inductive Reactance:** $X_L = 2\pi \times 60 \times 0.02 = 7.54$
  - **Impedance:** $Z = \sqrt{10^2 + 7.54^2} \approx 12.64$
  - **Current:** $I_{rms} = \frac{V_{rms}}{Z} = \frac{120}{12.64} \approx 9.49\ A$
  - **Real Power:** $P = V_{rms} I_{rms} cos\ (\phi)\ where\ cos\ (\phi) = \frac{R}{Z} = \frac{10}{12.64} \approx 0.79$

  $P = 120 \times 9.49 \times 0.79 \approx 899.56\ W$

Example 2

- A household electrical outlet supplies 240V AC at 50Hz. A capacitor with a capacitance of 100μF is connected across it.
  - **Capacitive Reactance:** $XC = \frac{1}{2\pi \times 50 \times 100 \times 10^{-6}} = 31.83$
  - **Current:** $I_{rms} = \frac{V_{rms}}{XC} = \frac{240}{31.83} \approx 7.54\ A$
  - **Reactive Power:** $Q = V_{rms} I_{rms} = 240 \times 7.54 \approx 1809.6\ VAR$

Applications

- **Household Electricity:** AC is used to power household appliances like refrigerators, air conditioners, and lights.
- **Power Transmission:** AC is preferred for transmitting electricity over long distances due to its ability to be easily transformed to different voltage levels.

- **Electronics:** AC circuits are used in various electronic devices, including radios, televisions, and computers, which often convert AC to DC for their internal operations.

9.2 Thevenin and Norton Theorems

Thevenin's Theorem is a method used in circuit analysis to simplify a complex linear electrical network into a simple equivalent circuit. This equivalent circuit consists of a single voltage source ($V_{Thevenin}$) in series with a single resistor ($R_{Thevenin}$). The theorem makes it easier to analyze the behavior of the circuit, especially when looking at the load's impact on the network.

Steps to Find Thevenin Equivalent Circuit

1. **Identify the portion of the circuit to be replaced with the Thevenin equivalent.**
   - This is usually the part where you need to find the load current or voltage.
2. **Remove the load resistor from the original circuit.**
3. **Find Thevenin Voltage ($V_{Thevenin}$):**
   - Calculate the open-circuit voltage across the terminals where the load resistor was connected.
   - This is the voltage that appears across the terminals with the load removed.
4. **Find Thevenin Resistance ($R_{Thevenin}$):**
   - Deactivate all independent voltage sources (replace them with short circuits) and independent current sources (replace them with open circuits).

- Calculate the equivalent resistance across the open terminals where the load resistor was connected.

5. **Draw the Thevenin Equivalent Circuit:**
   - Represent the circuit as a single voltage source ($V_{Thevenin}$) in series with a single resistor ($R_{Thevenin}$) connected to the load resistor.

Introduction to Norton's Theorem

Norton's Theorem is similar to Thevenin's Theorem but represents the circuit with a current source instead of a voltage source. The equivalent circuit consists of a single current source (I_{Norton}) in parallel with a single resistor (R_{Norton}). This theorem also simplifies the analysis of complex circuits, especially when considering the effect of varying the load.

Steps to Find Norton Equivalent Circuit

1. **Identify the portion of the circuit to be replaced with the Norton equivalent.**
   - This is usually the part where you need to find the load current or voltage.
2. **Remove the load resistor from the original circuit.**
3. **Find Norton Current (I_{Norton}):**
   - Calculate the short-circuit current across the terminals where the load resistor was connected.
   - This is the current that flows if the terminals are shorted together.

4. **Find Norton Resistance (R_{Norton}):**

   - Deactivate all independent voltage sources (replace them with short circuits) and independent current sources (replace them with open circuits).

   - Calculate the equivalent resistance across the open terminals where the load resistor was connected.

5. **Draw the Norton Equivalent Circuit:**

   - Represent the circuit as a single current source (I_{Norton}) in parallel with a single resistor (R_{Norton}) connected to the load resistor.

Examples and Applications

Example 1: Thevenin's Theorem

- **Original Circuit:** Consider a circuit with a 10V voltage source, a 2Ω resistor (R1), and a 4Ω resistor (R2) in series with a load resistor (R_L) of 6Ω.

   - **Step 1:** Identify the portion of the circuit to replace with the Thevenin equivalent.

   - **Step 2:** Remove the load resistor (R_L).

   - **Step 3:** Find $V_{Thevenin}$:

$$V_{Thevenin} = V_{R2} = V_{source} \times \frac{R2}{R1 + R2} = 10V \times \frac{4}{2 + 4} = 10V \times \frac{4}{6} = 6.67V$$

   - **Step 4:** Find $R_{Thevenin}$: $R_{Thevenin} = R1 + R2 = 2 + 4 = 6\ ohms$

   - **Step 5:** Draw the Thevenin equivalent circuit: A 6.67V source in series with a 6Ω resistor connected to the load resistor (R_L).

Example 2: Norton's Theorem

- **Original Circuit:** Consider the same circuit with a 10V voltage source, a 2Ω resistor (R1), and a 4Ω resistor (R2) in series with a load resistor (R_L) of 6Ω.
  - **Step 1:** Identify the portion of the circuit to replace with the Norton equivalent.
  - **Step 2:** Remove the load resistor (R_L).
  - **Step 3:** Find I_{Norton}: $I_{Norton} = \frac{V_{source}}{R1+R2} = \frac{10V}{2+4} = \frac{10V}{6} = 1.67A$
  - **Step 4:** Find R_{Norton} : $R_{Norton} = R1 + R2 = 2 + 4 = 6\ ohms$
  - **Step 5:** Draw the Norton equivalent circuit: A 1.67A current source in parallel with a 6Ω resistor connected to the load resistor (R_L).

Applications

- **Simplifying Complex Circuits:** Both Thevenin's and Norton's theorems help simplify complex circuits, making it easier to analyze and solve them.
- **Power Systems:** These theorems are used in power system analysis to determine the equivalent impedance and voltage or current at different points in the network.
- **Electronic Devices:** They are useful in designing and analyzing electronic circuits, such as amplifiers and signal processing circuits, where load effects need to be considered.

9.3 Waveform Analysis

Waveform analysis is essential in electrical engineering for understanding how circuits respond to different signals over time. By analyzing waveforms, engineers can predict circuit behavior, identify potential issues, and design more efficient systems. This section covers frequency response, transient response, and resonance in circuits, providing a comprehensive overview of how circuits react to various input signals.

Frequency Response

Frequency response describes how a circuit's output amplitude and phase change in relation to the frequency of the input signal. It is a crucial concept for designing and analyzing filters, amplifiers, and other signal processing circuits.

- **Gain (Magnitude Response):** Indicates how much the input signal is amplified or attenuated at different frequencies.
  - **Example:** A low-pass filter allows low-frequency signals to pass through while attenuating high-frequency signals.
- **Phase Response:** Indicates the phase shift introduced by the circuit at different frequencies.
  - **Example:** An audio equalizer adjusts the gain at various frequencies to enhance sound quality.

Example

- Consider an RC low-pass filter with a resistor (R) of 1kΩ and a capacitor (C) of 1μF.
  - **Cutoff Frequency (fc):** The frequency at which the output signal is attenuated to 70.7% of the input signal.

$$fc = \frac{1}{2\pi RC} = \frac{1}{2\pi \times 1000 \times 1 \times 10^{-6}} \approx 159.15\ Hz$$

  - At frequencies below 159.15 Hz, the output signal passes through with minimal attenuation. At frequencies above 159.15 Hz, the signal is significantly attenuated.

Transient Response

Transient response refers to how a circuit responds to a change in its input, such as a sudden application or removal of a voltage or current source. It is crucial for understanding the stability and dynamic behavior of circuits.

- **Step Response:** The circuit's reaction to a sudden change, like a switch turning on.
- **Impulse Response:** The circuit's reaction to a very short input signal (an impulse).

Example

- Consider an R_L circuit with a resistor (R) of 10Ω and an inductor (L) of 1H connected in series with a 5V DC source.
    - **Step Response:** $I(t) = \frac{V}{R}(1 - e^{-\left(\frac{R}{L}t\right)}) = \frac{5}{10}(1 - e^{-\frac{10}{1}t}) = 0.5(1 - e^{-10t})$
        - When the voltage is suddenly applied, the current in the circuit changes over time due to the inductor's opposition to changes in current.
        - The current starts at 0 and asymptotically approaches 0.5A.

Resonance in Circuits

Resonance occurs when a circuit is driven at its natural frequency, leading to a significant increase in voltage or current. It is particularly important in AC circuits involving inductors and capacitors.

- **Series Resonance:** In a series RLC circuit, resonance occurs when the inductive reactance (X_L) equals the capacitive reactance (X_C).

$$X_L = X_C \Rightarrow 2\pi f L = \frac{1}{2\pi f C} \Rightarrow fr = \frac{1}{2\pi\sqrt{LC}}$$

- **Parallel Resonance:** In a parallel RLC circuit, resonance occurs when the total impedance is maximized, and the current through the inductor and capacitor is equal and opposite, effectively canceling each other out.

Example

- Consider a series RLC circuit with a resistor (R) of 10Ω, an inductor (L) of 0.1H, and a capacitor (C) of 100μF.
  - **Resonant Frequency (f_r):**

$$f_r = \frac{1}{2\pi\sqrt{LC}} = \frac{1}{2\pi\sqrt{0.1 \times 100 \times 10^{-6}}} \approx 50.33\ Hz$$

  - At this frequency, the impedance of the inductor and capacitor cancel each other out, and the circuit's total impedance is purely resistive, leading to a maximum current.

Examples and Applications

Example 1

- **RC Low-Pass Filter:** An audio engineer uses an RC low-pass filter to remove high-frequency noise from a music signal.
  - **Components:** R = 1kΩ, C = 1μF.
  - **Cutoff Frequency:** 159.15 Hz.
  - The filter allows signals below 159.15 Hz to pass through, attenuating higher-frequency noise, resulting in a cleaner audio output.

Example 2

- **RL Circuit Transient Response:** An electric motor control system uses an RL circuit to manage the startup current of the motor.
  - **Components:** R = 10Ω, L = 1H, V = 5V DC.

- **Current Response:** The current starts at 0 and gradually increases to 0.5A, preventing a sudden surge that could damage the motor.

Example 3

- **Series RLC Circuit at Resonance:** A radio tuner uses a series RLC circuit to select a specific frequency from a range of broadcast signals.
  - **Components:** R = 10Ω, L = 0.1H, C = 100μF.
  - **Resonant Frequency:** 50.33 Hz.
  - The tuner adjusts the inductance or capacitance to resonate at the desired frequency, maximizing the signal reception while minimizing interference.

Applications

- **Filters:** Low-pass, high-pass, band-pass, and band-stop filters use frequency response principles to allow or block specific frequency ranges.
- **Oscillators:** Circuits designed to produce periodic signals use resonance to maintain stable frequencies.
- **Signal Processing:** Transient response analysis helps design circuits that respond predictably to changing signals, such as pulse circuits in digital electronics.

9.4 Capacitance and Inductance

Capacitance is the ability of a component to store electrical energy in an electric field. It is primarily used in electronic circuits to store charge, filter signals, and manage power supply stability. The unit of capacitance is the farad (F), which measures the amount of charge a capacitor can store per volt of potential difference.

- **Capacitor:** A device consisting of two conductive plates separated by an insulating material (dielectric). When a voltage is applied across the plates, an electric field develops, and charge is stored.

Formula: The capacitance (C) is given by: $C = \frac{Q}{V}$ where Q is the charge stored, and V is the voltage across the capacitor.

Example

- A capacitor with a capacitance of 10μF connected to a 12V battery stores:

$$Q = C \times V = 10 \times 10^{-6} \times 12 = 120\mu C$$

Capacitors in Series and Parallel

Capacitors in Series

- When capacitors are connected in series, the total capacitance (C_{total}) decreases.

$\frac{1}{C_{total}} = \frac{1}{C_1} + \frac{1}{C_2} + \ldots + \frac{1}{C_n}$

- Two capacitors, 10μF and 20μF, connected in series:

$$\frac{1}{C_{total}} = \frac{1}{10} + \frac{1}{20} = \frac{2}{20} + \frac{1}{20} = \frac{3}{20}$$

$$C_{total} = \frac{20}{3} \approx 6.67\mu F$$

Capacitors in Parallel

- When capacitors are connected in parallel, the total capacitance (C_{total}) increases.

$$C_{total} = C_1 + C_2 + \cdots + C_n$$

- Two capacitors, 10μF and 20μF, connected in parallel:

$$C_{total} = 10 + 20 = 30\mu F$$

Introduction to Inductance

Inductance is the property of a component that opposes changes in current flow by storing energy in a magnetic field. Inductors are used in circuits for filtering, energy storage, and in applications requiring controlled current changes. The unit of inductance is the Henry (H).

- **Inductor:** A coil of wire that generates a magnetic field when current flows through it. The inductance (L) depends on the number of turns in the coil, the cross-sectional area, and the core material.

Formula: The inductance (L) is given by: $L = \frac{N^2 \mu A}{l}$ where N is the number of turns, μ is the permeability of the core material, A is the cross-sectional area, and l is the length of the coil.

Inductors in Series

- When inductors are connected in series, the total inductance (L_{total}) increases.

$$L_{total} = L_1 + L_2 + \cdots + L_n$$

- Two inductors, 3H and 6H, connected in series:

$$L_{total} = 3 + 6 = 9\ H$$

Inductors in Parallel

- When inductors are connected in parallel, the total inductance (L_{total}) decreases.

$$\frac{1}{L_{total}} = \frac{1}{L_1} + \frac{1}{L_2} + \cdots + \frac{1}{L_n}$$

- Two inductors, 3H and 6H, connected in parallel:

$$\frac{1}{L_{total}} = \frac{1}{3} + \frac{1}{6} = \frac{2}{6} + \frac{1}{6} = \frac{3}{6}$$

$$L_{total} = \frac{6}{3} = 2\ H$$

Energy Stored in Capacitors and Inductors

Energy in Capacitors

- The energy (W) stored in a capacitor is given by: $W = \frac{1}{2} CV^2$

- A 10μF capacitor charged to 12V stores:

$$W = \frac{1}{2} \times 10 \times 10^{-6} \times 12^2 = 720\mu J$$

Energy in Inductors

- The energy (W) stored in an inductor is given by:

$$W = \frac{1}{2} LI^2$$

- A 2H inductor with a current of 5A stores:

$$W = \frac{1}{2} \times 2 \times 5^2 = 25\,J$$

Examples and Applications

Example 1: Filtering Noise

- **Application:** Capacitors and inductors are used in filter circuits to remove unwanted noise from signals.
- **Circuit:** An RC low-pass filter uses a capacitor in parallel with the signal path to filter out high-frequency noise.

Example 2: Inductive Kickback

- **Application:** Inductors are used in circuits to manage inductive kickback, a surge of voltage when the current through an inductor is suddenly interrupted.
- **Circuit:** In a relay circuit, a diode is placed across the inductor coil to protect other components from high voltage spikes caused by inductive kickback.

Example 3: Energy Storage in Power Supplies

- **Application:** Capacitors store energy in power supplies to provide a smooth and stable voltage output.

- **Circuit:** In a power supply circuit, large capacitors are used to filter out voltage ripples and maintain a steady DC output.

Example 4: Resonant Circuits

- **Application:** LC circuits (combining inductors and capacitors) are used in tuning and resonance applications, such as radio transmitters and receivers.
- **Circuit:** An LC circuit in a radio tuner selects a specific frequency from a range of broadcast signals by resonating at that frequency.

Applications

- **Signal Processing:** Capacitors and inductors are essential in designing filters that allow or block specific frequency ranges, critical in audio and communication systems.
- **Power Systems:** Inductors are used in transformers to transfer energy between circuits through magnetic fields, essential in power distribution.
- **Energy Storage:** Capacitors provide quick bursts of energy in applications like camera flashes and power supplies, ensuring consistent performance.
- **Timing Circuits:** RC and LC circuits create precise time delays and oscillations, vital in clocks, timers, and frequency generators.

9.5 Power Calculations

Power in DC Circuits

Power in direct current (DC) circuits is straightforward to calculate. It is the product of voltage and current, representing the rate at which energy is consumed or produced.

- **Formula:** $P = V \times I$

where P is power in watts (W), V is voltage in volts (V), and I is current in amperes (A).

Example

- A light bulb operates with a voltage of 12V and a current of 2A.

$$P = 12V \times 2A = 24W$$

The light bulb consumes 24 watts of power.

Power in AC Circuits

Power calculations in alternating current (AC) circuits are more complex due to the changing nature of the voltage and current. Power can be divided into real power, reactive power, and apparent power.

- **Real Power (P):** The actual power consumed by the resistive components of the circuit, measured in watts (W).
- **Reactive Power (Q):** The power stored and released by the inductive and capacitive components, measured in volt-amperes reactive (VAR).
- **Apparent Power (S):** The combined effect of real and reactive power, measured in volt-amperes (VA).

Formulas

- **Real Power:**

$$P = V_{rms} \times I_{rms} \times cos(\phi)$$

  - Where ϕ is the phase angle between voltage and current.
- **Reactive Power:**

$$Q = V_{rms} \times I_{rms} \times sin(\phi)$$

- **Apparent Power:** $S = V_{rms} \times I_{rms}$

Example

- An AC circuit with $V_{rms} = 120V, I_{rms} = 10A, and\ \phi = 30^{\circ}$

$$P = 120V \times 10A \times cos(30^{\circ}) = 1200 \times 0.866 = 1039.2W$$

$$Q = 120V \times 10A \times sin(30^\circ) = 1200 \times 0.5 = 600VAR$$

$$S = 120V \times 10A = 1200VA$$

Power Factor and its Importance

The power factor is the ratio of real power to apparent power in an AC circuit. It indicates how effectively electrical power is being used.

Formula

$$Power\ Factor = cos(\phi) = \frac{P}{S}$$

Importance

- **Efficiency:** A higher power factor means more efficient use of electrical power. A lower power factor indicates poor efficiency and results in higher power losses.
- **Cost:** Utilities may charge higher rates for industrial customers with low power factors because it indicates inefficient use of the power system.

Example

- In the previous example, the power factor is:

$$Power\ Factor = cos(30^\circ) = 0.866$$

Real, Reactive, and Apparent Power

- **Real Power (P):** The power that performs actual work in the circuit, such as lighting a bulb or running a motor.
- **Reactive Power (Q):** The power that oscillates between the source and reactive components (inductors and capacitors) without doing any actual work.
- **Apparent Power (S):** The vector sum of real and reactive power, representing the total power supplied by the source.

Formulas

- **Real Power:** $P = V_{rms} \times I_{rms} \times \cos(\phi)$

- **Reactive Power:** $Q = V_{rms} \times I_{rms} \times \sin(\phi)$

- **Apparent Power:** $S = \sqrt{P^2 + Q^2}$

Example

- For an AC circuit with real power P=800W and reactive power Q=600V AR:

$$S = \sqrt{800^2 + 600^2} = \sqrt{640000 + 360000} = \sqrt{1000000} = 1000V\ A$$

Examples and Applications

Example 1

- **DC Circuit Power Calculation:**
    - A DC motor operates at 24V and 5A.

    $$P = 24V \times 5A = 120W$$

    - The motor consumes 120 watts of power.

Example 2

- **AC Circuit Power Calculation:**
    - An AC circuit has a voltage of 230V, a current of 10A, and a power factor of 0.8.

    $$P = 230V \times 10A \times 0.8 = 1840W$$

    $$Q = 230V \times 10A \times \sin(\cos^{-1}(0.8)) = 2300 \times 0.6 = 1380V\ AR$$

    $$S = 230V \times 10A = 2300V\ A$$

Applications

- **Industrial Equipment:** Understanding power calculations helps design efficient machinery and manage power consumption effectively.

- **Power Grids:** Utilities use these calculations to ensure efficient power distribution and minimize losses.

- **Renewable Energy Systems:** Engineers use power calculations to optimize the performance of solar panels and wind turbines, ensuring maximum energy output and storage efficiency.

ADDITIONAL ONLINE RESOURCES

- **1500+ PRACTICE PROBLEMS [4 PRACTICE EXAMS INCLUDED]**

Enhance your exam preparation with over 1000+ practice problems, complete with step-by-step solutions. Two weeks after your initial download, you will receive a weekly full-length practice exam with step-by-step solutions for four weeks, totaling over 1500 practice problems.

SCAN HERE

Having trouble accessing the resources? Email us at **info@learnovapublishing.com** — we reply within 24 hours, 7 days a week

- **VIDEO RESOURCE RECOMMENDATIONS**

Enhance your learning with our recommended video resources for a deeper understanding of the concepts. When using these resources, please check the current exam specifications.

An abundance of high-quality video resources is available for preparing for this section. For a comprehensive overview, we recommend the following:

YouTube Playlist: *FE Electrical and Computer: Review (Effective from July 2020) by Amit Roy*
Search for this playlist on YouTube. It covers almost all the exam topics, providing detailed explanations and condensing the material effectively.

Online Review Course: For a more organized approach, consider the comprehensive, free, video-based online review course called *"Electrical FE Review"* by Zach Stone, available at electricalfereview.com

Ohm's Law and Basic Principles	
Key Concepts	**Handbook Page References**
Ohm's Law	
Electrical Power	
Kirchhoff's Voltage Law (KVL)	
Kirchhoff's Current Law (KCL)	

Circuit Arrangements and Components	
Key Concepts	**Handbook Page References**
Series and Parallel Circuits	
Open Circuit	
Short Circuit	
Multiple Sources (Superposition Theorem)	
Inductors:	
- Characteristics	
- Inductors in Parallel and in Series	
Capacitors:	
- Characteristics	
- Capacitors in Parallel and in Series	

Thevenin and Norton Theorems	
Key Concepts	**Handbook Page References**
Thevenin Equivalent	
Norton Equivalent	
Source Transformation	

Node and Loop Analysis	
Key Concepts	**Handbook Page References**
Node Analysis	
Mesh Analysis	

Waveform Analysis	
Key Concepts	**Handbook Page References**
Frequency	
Average Value	
RMS Value	
Maximum Value	

Phasors and Impedance	
Key Concepts	**Handbook Page References**
Phasors:	
- Rectangular Form	
- Polar Form	
- Conversions between Rectangular and Polar Forms	
Impedance:	
- Resistance (Resistors	

- Inductance or Inductive Reactance (Inductors)	
- Capacitance or Capacitive Reactance (Capacitors)	

Frequency/Transient Response	
Key Concepts	**Handbook Page References**
RL Transient Response	
RL Frequency Response:	
- High Pass and Low Pass Filters	
RC Transient Response	
RC Frequency Response:	
- High Pass and Low Pass Filters	

Resonance	
Key Concepts	**Handbook Page References**
Series RLC Resonance:	
- Series RLC Resonant Frequency	
- Series RLC Bandwidth	
Parallel RLC Resonance:	
- Parallel RLC Resonant Frequency	
- Parallel RLC Bandwidth	
- Parallel RLC Bandwidth - Current Source	

Laplace Transforms	
Key Concepts	**Handbook Page References**
Laplace & Z Transform Pairs	

Laplace Properties:	
- Linearity	
- First, Second & Nth Derivative	
- Integral	
- Time Multiplication/Division by Time	
- Time Delay/Advancement	
- Complex Shift	
- Scale	
- Convolution	

Transfer Functions	
Key Concepts	**Handbook Page References**
Poles & Zeroes	
Two-Port Network:	
- Impedance Parameters	

NOTES/EQUATIONS SUMMARY

DIAGNOSTIC TEST CHECKLIST

Checklist Questions	Yes/No
Have you read and understood all the material in this chapter?	
Are you familiar with all the key concepts and formulas highlighted in the tables for this chapter?	
Have you filled in the tables with the correct NCEES® FE Reference Handbook page numbers for each listed concept and formula?	
Did you download and complete all the designated practice problems for this section?	
Did you review and understand the step-by-step solutions to the practice problems?	
Could you solve the practice problems without excessive reference to the solutions and within a reasonable time frame?	
Are you confident in applying the concepts and formulas from this chapter?	
Have you noted any areas of difficulty or topics that require further review before proceeding?	
Do you know how to quickly find information in the NCEES® FE Reference Handbook?	
Have you reviewed any errors or misunderstandings in this section's material to ensure clarity and understanding?	

CHAPTER 10

Properties of Electrical Materials and Electronics

Understanding the properties of electrical materials and the fundamentals of electronics is crucial for any electrical engineer. These properties determine how materials behave in different electrical and electronic applications, influencing the design, functionality, and efficiency of circuits and devices.

This chapter delves into the essential properties of electrical materials, such as conductivity, resistivity, permittivity, permeability, capacitance, inductance, and impedance. These properties form the foundation for analyzing and selecting materials for specific applications, ensuring that the materials meet the necessary electrical and mechanical requirements.

10.1 Properties of Electrical Materials

Conductivity

Definition and Significance

- Conductivity is the measure of a material's ability to conduct electric current. It is the inverse of resistivity and is measured in siemens per meter (S/m).
- Conductive materials, like copper and aluminum, are used in electrical wiring and components because they allow electric current to flow with minimal resistance.

Factors Affecting Conductivity

- **Material Composition:** Pure metals have high conductivity, while alloys may have lower conductivity.
- **Temperature:** Conductivity decreases as temperature increases in most conductive materials due to increased atomic vibrations that interfere with electron flow.
- **Impurities:** The presence of impurities in a material can decrease its conductivity by scattering electrons.

Example

- Copper, widely used in electrical wiring, has a high conductivity of approximately $5.96 \times 10^7\ S/m$, making it an excellent conductor of electricity.

Resistivity

Definition and Importance

- Resistivity is the measure of a material's resistance to the flow of electric current, measured in ohm-meters (Ω·m). It indicates how strongly a material opposes the flow of electric current.

- Materials with high resistivity, such as rubber and glass, are used as insulators.

Relationship with Conductivity

- Conductivity and resistivity are inversely related: $\sigma = \frac{1}{\rho}$

  where σ is conductivity and ρ is resistivity.

Factors Affecting Resistivity

- **Material Composition:** Different materials have inherently different resistivities.

- **Temperature:** For most conductive materials, resistivity increases with temperature.

- **Impurities and Defects:** Impurities and structural defects in a material can increase its resistivity.

Example

- Nichrome, an alloy used in heating elements, has a resistivity of about 1.10×10^{-6} Ω·m, which is higher than that of pure metals like copper.

Permittivity

Definition and Role in Capacitors

- Permittivity is a measure of how easily a material can store electrical energy in an electric field. It is essential in the design and function of capacitors.

- Permittivity determines how much electric charge a capacitor can store for a given electric field strength.

Types of Permittivity

- **Absolute Permittivity (ε):** The measure of a material's ability to permit electric field lines.

- **Relative Permittivity (ε_r):** The ratio of the permittivity of a material to the permittivity of a vacuum ($\boldsymbol{\varepsilon_0}$): $\epsilon_r = \frac{\epsilon}{\epsilon_0}$

Example

- The relative permittivity of a dielectric material used in a capacitor can significantly affect its capacitance. For example, ceramic capacitors use materials with high relative permittivity to achieve higher capacitance values.

Permeability

Definition and Importance in Inductors

- Permeability is the measure of a material's ability to support the formation of a magnetic field within itself. It is crucial in the design of inductors and transformers.

- Materials with high permeability are used in the cores of inductors and transformers to enhance their inductance and efficiency.

Types of Permeability

- **Absolute Permeability (μ):** The measure of a material's ability to conduct a magnetic field.

- **Relative Permeability (μ_r):** The ratio of the permeability of a material to the permeability of free space ($\boldsymbol{\mu_0}$): $\mu_r = \frac{\mu}{\mu_0}$

Example

- Ferromagnetic materials like iron have high relative permeability and are commonly used in the cores of inductors and transformers to increase their inductance.

Capacitance

Definition and Role in Circuits

- Capacitance is the ability of a component to store electrical energy in the form of an electric field. It is measured in farads (F).
- Capacitors are used in circuits to store energy, filter signals, and stabilize power supply voltages.

Factors Affecting Capacitance

- **Surface Area of Plates:** Larger plate areas result in higher capacitance.
- **Distance Between Plates:** Smaller distances between plates increase the capacitance.
- **Dielectric Material:** Materials with higher permittivity between the plates increase the capacitance.

Example

- A parallel plate capacitor with plates of area $1\,m^2$ separated by 1 mm and a dielectric material with a relative permittivity of 5 will have a higher capacitance than the same capacitor with air as the dielectric.

Inductance

Definition and Role in Circuits

- Inductance is the property of a conductor by which a change in current flowing through it induces an electromotive force (EMF). It is measured in henries (H).
- Inductors are used in circuits to store energy in magnetic fields, filter signals, and manage current changes.

Factors Affecting Inductance

- **Number of Turns:** More turns in the coil result in higher inductance.
- **Core Material:** Materials with higher permeability increase inductance.
- **Cross-sectional Area:** Larger cross-sectional areas of the coil increase inductance.
- **Length of the Coil:** Shorter coils have higher inductance.

Example

- An inductor with 100 turns, a core of high-permeability material, and a larger cross-sectional area will have a higher inductance than a similar inductor with fewer turns and a smaller core.

Impedance

Definition and Significance in AC Circuits

- Impedance is the measure of opposition that a circuit presents to the passage of alternating current (AC). It is measured in ohms (Ω).
- Impedance combines both resistance (R) and reactance (X), which includes inductive reactance (X_L) and capacitive reactance (X_C).

Components of Impedance

- **Resistance (R):** The opposition to current flow due to resistive components.
- **Reactance (X):** The opposition to current flow due to inductive and capacitive components.

$$Z = \sqrt{R^2 + X^2}$$

$$X = X_L - X_C$$

$$X_L = 2\pi f L$$

$$X_C = \frac{1}{2\pi f C}$$

Example

- An AC circuit with a resistor of 10Ω and an inductor with an inductive reactance of 15Ω has an impedance of:

$$Z = \sqrt{10^2 + 15^2} = \sqrt{100 + 225} = \sqrt{325} \approx 18\Omega$$

10.2 Semiconductors

Definition and Types

Semiconductors are materials that have electrical conductivity between that of conductors (like metals) and insulators (like glass). They are the backbone of modern electronics, enabling the function of various devices such as diodes, transistors, and integrated circuits. There are two main types of semiconductors:

- **Intrinsic Semiconductors:** Pure semiconductor materials without any impurities. Silicon and germanium are common examples.
- **Extrinsic Semiconductors:** Semiconductor materials that have been intentionally doped with impurities to modify their electrical properties.

Example

- Silicon, a widely used semiconductor, can act as an insulator at low temperatures and as a conductor at higher temperatures due to its ability to conduct electricity only under certain conditions.

Doping and Carrier Concentration

Process of Doping

Doping is the process of adding impurity atoms to a semiconductor to change its electrical properties. This process increases the number of charge carriers (electrons or holes) in the material, enhancing its conductivity.

- **n-Type Doping:** Adding donor impurities (like phosphorus or arsenic) that have more valence electrons than the semiconductor. This increases the number of free electrons.

- **p-Type Doping:** Adding acceptor impurities (like boron or gallium) that have fewer valence electrons than the semiconductor. This creates more holes (positive charge carriers).

Effects on Electrical Properties

Doping significantly affects the electrical properties of semiconductors:

- **Increased Conductivity:** Doped semiconductors have higher conductivity due to the increased number of charge carriers.

- **Controlled Electrical Behavior:** By controlling the type and amount of doping, manufacturers can tailor semiconductors to specific applications.

Example

- Doping silicon with phosphorus atoms creates an n-type semiconductor with extra free electrons while doping it with boron creates a p-type semiconductor with extra holes.

PN Junction and Diodes

Formation of PN Junction

A PN junction is formed by joining p-type and n-type semiconductor materials. This junction is the basic building block of many semiconductor devices, including diodes.

- **Depletion Region:** At the junction, electrons from the n-type region combine with holes from the p-type region, creating a depletion region with no free charge carriers.

Working Principle of Diodes

A diode allows current to flow in one direction only, making it a crucial component in rectifying circuits.

- **Forward Bias:** When the p-type side is connected to the positive terminal of a battery and the n-type side to the negative terminal, the diode conducts current.
- **Reverse Bias:** When the connections are reversed, the diode blocks current flow.

Example

- In a silicon diode, when forward biased, the depletion region narrows, allowing current to flow easily. In reverse bias, the depletion region widens, preventing current flow.

Applications of Semiconductors

Semiconductors are integral to numerous electronic devices due to their versatile properties:

- **Transistors:** Used as amplifiers and switches in various electronic circuits.
- **Integrated Circuits (ICs):** Comprise numerous transistors and other components on a single chip, used in computers, smartphones, and other digital devices.
- **Solar Cells:** Convert light energy into electrical energy using semiconductor materials.
- **LEDs:** Light-emitting diodes that produce light when current flows through them.

Example

- A transistor made from silicon can act as a switch to turn on and off the current in a circuit, making it essential for digital computing.

10.3 Operational Amplifiers

Definition and Basic Functions

Operational amplifiers, or op-amps, are versatile and widely used electronic components that amplify voltage. They are essential in various analog electronic circuits due to their high gain, high input impedance, and low output impedance.

Basic Functions

- **Amplification:** Increasing the amplitude of a signal.
- **Signal Conditioning:** Modifying a signal to meet the requirements of the next stage in a system.
- **Mathematical Operations:** Performing operations such as addition, subtraction, integration, and differentiation.

Example

- An op-amp can amplify a small sensor signal to a level suitable for processing by a microcontroller.

Ideal vs. Real Op-Amps

Key Differences

- **Ideal Op-Amps:** These are theoretical models used for simplifying circuit analysis.
    - **Infinite Gain:** The open-loop gain (A) is infinite.
    - **Infinite Input Impedance:** No current flows into the input terminals.

- **Zero Output Impedance:** The output can drive any load without voltage drop.
- **Infinite Bandwidth:** It can amplify any frequency without attenuation.
- **Zero Offset Voltage:** The output is zero when the input is zero.

- **Real Op-Amps:** These have practical limitations.
  - **Finite Gain:** Typically ranges from 10^5 *to* 10^7.
  - **High but Finite Input Impedance:** Ranges from 1MΩ to 10MΩ.
  - **Low but Non-zero Output Impedance:** Typically, a few ohms.
  - **Limited Bandwidth:** Gain decreases with increasing frequency.
  - **Offset Voltage:** A small voltage present at the output when the input is zero.

Example

- Due to bandwidth limitations, a real op-amp with a gain of 100,000 may not amplify a very high-frequency signal as efficiently as it does a low-frequency signal.

Common Configurations

Inverting Amplifier

- **Configuration:** The input signal is applied to the inverting input (-) through a resistor, while the non-inverting input (+) is grounded. Feedback is provided from the output to the inverting input through another resistor.
- **Gain:** The voltage gain (A) is determined by the ratio of the feedback resistor (R_f) to the input resistor (R_{in})

$$A = -\frac{R_f}{R_{in}}$$

- **Phase:** The output signal is 180 degrees out of phase with the input signal.

- If $R_f = 10k\Omega$ and $R_{in} = 1k\Omega$, the gain is: $A = -\frac{10k\Omega}{1k\Omega} = -10$

Non-Inverting Amplifier

- **Configuration:** The input signal is applied to the non-inverting input (+), while the inverting input (-) is connected to a voltage divider made up of two resistors, with feedback from the output.

- **Gain:** The voltage gain (A) is given by: $A = 1 + \frac{R_f}{R_{in}}$

- **Phase:** The output signal is in phase with the input signal.

- If $R_f = 10k\Omega \; and \; R_{in} = 2k\Omega$, the gain is: $A = 1 + \frac{10k\Omega}{2k\Omega} = 6$

Differential Amplifier

- **Configuration:** The input signals are applied to both the inverting and non-inverting inputs. The output voltage is proportional to the difference between the input voltages.

- **Gain:** The voltage gain (A) is determined by the ratio of resistors used in the circuit: $A = \frac{R_f}{R_{in}}$

- **Phase:** The output signal is proportional to the difference between the input signals.

- If both resistors R_f and R_{in} are 10kΩ, the gain is: $A = \frac{10k\Omega}{10k\Omega} = 1$

Applications of Operational Amplifiers

Signal Conditioning

- Op-amps are used to amplify, filter, and buffer signals in sensor interfaces to make them compatible with data acquisition systems.

Filtering

- Op-amps are essential in designing active filters (low-pass, high-pass, band-pass, and band-stop) for processing audio signals, removing noise, and enhancing specific frequency components.

Mathematical Operations

- They are used in analog computers to perform operations such as integration, differentiation, summation, and subtraction, which are crucial in control systems and signal processing.

Voltage Follower (Buffer)

- Op-amps can be configured as voltage followers to provide high input impedance and low output impedance, useful in impedance matching.

Example

- In an audio amplifier circuit, op-amps amplify weak audio signals from a microphone to drive speakers with higher power.

10.4 Digital Electronics

Definition and Significance

Digital electronics involves circuits and systems that use digital signals to represent and process information. Unlike analog signals, which vary continuously, digital signals have discrete levels, usually represented as binary numbers (0s and 1s). This binary representation forms the foundation of modern computing and digital communication technologies.

Significance

- Digital electronics enables the design and operation of computers, smartphones, and other digital devices.
- It allows for precise and error-resistant data processing and storage.

Example

- The digital clock in a smartphone processes binary data to display the correct time accurately.

Logic Gates and Boolean Algebra

Types of Logic Gates

Logic gates are the basic building blocks of digital circuits. Each gate performs a specific logical function based on Boolean algebra.

- **AND Gate:** Outputs true (1) only if all inputs are true.
- **OR Gate:** Outputs true if at least one input is true.
- **NOT Gate (Inverter):** Outputs the opposite of the input.
- **NAND Gate:** Outputs false (0) only if all inputs are true.
- **NOR Gate:** Outputs true only if all inputs are false.
- **XOR Gate:** Outputs true if the inputs are different.
- **XNOR Gate:** Outputs true if the inputs are the same.

Basic Boolean Operations

Boolean algebra is a mathematical framework used to describe the relationships between binary variables and logical operations.

- **AND Operation:** $A \cdot B$

- **OR Operation:** $A + B$
- **NOT Operation:** $\bar{A}$
- **NAND Operation:** $\overline{A \cdot B}$
- **NOR Operation:** $\overline{A + B}$
- **XOR Operation:** $A \oplus B$
- **XNOR Operation:** $\overline{A \oplus B}$

Example

- A logic circuit with two inputs (A and B) and an AND gate outputs 1 only when both A and B are 1.

Combinational and Sequential Circuits

Definitions and Examples

- **Combinational Circuits:** These circuits generate an output based solely on the current inputs, without any memory element.
  - **Examples:** Adders, multiplexers, and decoders.
  - **Example Circuit:** A half-adder that adds two single-bit binary numbers (A and B) produces a sum (S) and carries (C) output.

    $S = A \oplus B \text{ and } C = A \cdot B$
- **Sequential Circuits:** These circuits generate an output based on both current inputs and previous states, using memory elements like flip-flops.
  - **Examples:** Counters, shift registers, and flip-flops.
  - **Example Circuit:** A D flip-flop that stores the value of the input (D) when a clock pulse is received: $Q_{next} = D$

Microcontrollers and Microprocessors

Microcontrollers

These are integrated circuits designed to perform specific control tasks in embedded systems. They contain a processor, memory, and input/output peripherals on a single chip.

- **Applications:** Used in household appliances, automotive systems, medical devices, and more.
- **Example:** An Arduino microcontroller can control the temperature of a thermostat based on sensor input.

Microprocessors

These are the central processing units (CPUs) of computers, designed to execute general-purpose computing tasks. They contain the arithmetic logic unit (ALU), control unit, and registers.

- **Applications:** Used in personal computers, servers, and smartphones.
- **Example:** An Intel Core i7 microprocessor executes complex instructions to run applications, manage memory, and handle input/output operations.

Example

- A microcontroller in a washing machine controls the washing cycles, water levels, and spin speed based on user inputs and sensor data.

10.5 Circuit Design and Analysis

Importance and Methodologies

Circuit design is the process of creating electrical circuits that perform specific functions. This process is crucial for developing everything from simple electronic gadgets to complex communication systems. Effective circuit design ensures the reliability, efficiency, and optimal performance of electronic devices.

Importance

- Ensures that circuits function correctly and meet design specifications.
- Minimizes production costs by optimizing component selection and layout.
- Enhances the reliability and lifespan of electronic devices.

Methodologies

- **Specification Definition:** Clearly defining the requirements and constraints of the circuit.
- **Schematic Design:** Creating a detailed diagram that represents the electrical connections and components.
- **PCB Layout:** Designing the physical layout of the circuit on a printed circuit board (PCB).
- **Testing and Validation:** Ensuring that the circuit meets all specifications and functions as intended.

Example

- Designing a power supply circuit involves defining voltage and current requirements, selecting appropriate components, and creating a schematic that ensures stable output.

Design Techniques

Top-down and Bottom-up Approaches

Two primary design approaches are used in circuit design: top-down and bottom-up.

- **Top-down Approach**
    - Starts with the overall system design and breaks it down into smaller sub-circuits.
    - Focuses on system-level requirements and ensures that all components work together seamlessly.
    - **Example:** Designing a microcontroller-based system by first defining the overall functionality and then designing each module (e.g., power supply, input/output interfaces) separately.
- **Bottom-up Approach**
    - Begins with designing individual components or modules and integrates them into a complete system.
    - Focuses on optimizing each part of the circuit for performance and efficiency.
    - **Example:** Developing a high-performance amplifier by first designing and testing individual amplifier stages before integrating them into a complete system.

Example

- In a top-down design of a smart thermostat, the overall functionality is defined first, followed by the design of specific modules like temperature sensors, display units, and control circuits.

Simulation and Prototyping

Tools and Processes

Simulation and prototyping are essential steps in the circuit design process, allowing designers to test and refine their designs before production.

- **Simulation**
    - Uses software tools to model and analyze the behavior of the circuit.
    - Identifies potential issues and optimizes performance without the need for physical components.
    - **Tools:** SPICE (Simulation Program with Integrated Circuit Emphasis), Multisim, and LTspice.
    - **Process:** Creating a virtual model of the circuit, running simulations to test various scenarios, and analyzing results to make necessary adjustments.
- **Prototyping**
    - Involves building a physical version of the circuit to test its functionality in real-world conditions.
    - Allows designers to verify the performance and make final adjustments.
    - **Tools:** Breadboards, PCB prototypes, and development kits.
    - **Process:** Assembling the circuit on a breadboard or custom PCB, conducting tests, and refining the design based on test results.

Example

- Simulating an audio amplifier circuit using LTspice to ensure it meets gain and frequency response requirements, followed by building a prototype to test its performance with actual audio signals.

Troubleshooting and Optimization

Common Issues and Solutions

Troubleshooting involves identifying and resolving problems that arise during the design and testing phases.

- **Common Issues**
  - **Incorrect Component Values:** Ensure components match the design specifications.
  - **Signal Interference:** Minimize noise and interference through proper shielding and layout techniques.
  - **Power Supply Problems:** Verify voltage levels and stability of the power supply.
- **Solutions**
  - **Component Testing:** Use multimeters and oscilloscopes to test individual components and connections.
  - **Layout Adjustments:** Improve PCB layout to reduce interference and enhance signal integrity.
  - **Thermal Management:** Implement heat sinks or cooling systems to manage heat dissipation.

Example

- If an amplifier circuit exhibits unexpected noise, using an oscilloscope to trace the source of the noise and adjusting the layout or adding filtering components can resolve the issue.

Techniques for Optimizing Circuit Performance

Optimization involves enhancing the circuit's performance to meet or exceed design specifications.

- **Techniques**
    - **Component Selection:** Choose high-quality components with suitable specifications for the application.
    - **Signal Integrity**: Ensure clean signal paths by minimizing trace lengths and using proper grounding techniques.
    - **Thermal Management:** Design for efficient heat dissipation to maintain stable operation.
    - **Power Efficiency:** Optimize power consumption by selecting energy-efficient components and designing low-power circuits.

Example

- Optimizing a battery-powered device by selecting low-power components and implementing power-saving modes to extend battery life.

ADDITIONAL ONLINE RESOURCES

- **1500+ PRACTICE PROBLEMS [4 PRACTICE EXAMS INCLUDED]**

Enhance your exam preparation with over 1000+ practice problems, complete with step-by-step solutions. Two weeks after your initial download, you will receive a weekly full-length practice exam with step-by-step solutions for four weeks, totaling over 1500 practice problems.

SCAN HERE

Having trouble accessing the resources? Email us at **info@learnovapublishing.com** — we reply within 24 hours, 7 days a week

- **VIDEO RESOURCE RECOMMENDATIONS**

Enhance your learning with our recommended video resources for a deeper understanding of the concepts. When using these resources, please check the current exam specifications.

An abundance of high-quality video resources is available for preparing for this section. For a comprehensive overview, we recommend the following:

YouTube Playlist: *Properties of Electrical Materials (NCEES® Electrical and Computer FE Exam) by Electrical FE Review*

Search for this playlist on YouTube. It provides detailed explanations condensing the material effectively.

Online Review Course: For a more organized approach, consider the comprehensive, free, video-based online review course called *"Electrical FE Review"* by Zach Stone, available at electricalfereview.com

Semiconductor Materials	
Key Concepts	**Handbook Page References**
Electrical Carrier Concentrations:	
- Equilibrium Carrier Concentration	
- Doped p-type Material	
- Doped n-type Material	
- Conductivity of a Semiconductor Material	
- Resistance of a Diffused Layer	
Diffusion of Defects:	
- Diffusion Coefficient	
Tunneling	
Energy Bands:	
- Band Theory	
- Energy Band Diagrams	

Electrical Properties	
Key Concepts	**Handbook Page References**
Conductivity	
Resistivity and Resistance	
Permittivity:	
- Capacitors	
- Transmission Lines	
Temperature Coefficient	
Magnetic Properties:	
- Reluctance	
Permeability:	

- Inductors	
Capacitance:	
- Capacitors	
- Capacitors in Series and Parallel	
- Ideal Capacitor	
Inductance:	
- Inductors	
- Ideal Inductor	
Impedance:	
- Impedance and Ohm's Law	
- Ideal Resistor	

Thermal Properties	
Key Concepts	**Handbook Page References**
Resistance in Series	

Modes, Biasing, and Performance of Discrete Devices	
Key Concepts	**Handbook Page References**
Diodes:	
- Diodes Biasing	
- Thermal Voltage	
- Shockley's Equation	
- Zener Diode	
Thyristors	
Insulated Gate Bipolar Transistors (IGBTs)	
Bipolar Junction Transistor (BJT):	
- Common Base Configuration	

- Common Emitter Configuration	
- Common Collector Configuration	
- BJT Terminals Mathematical Relations	

JFETs and MOSFETs	
Key Concepts	**Handbook Page References**
Junction Field Effect Transistor (JFET):	
- N-Channel JFET	
- P-Channel JFET	
- JFET Mathematical Relationships	
Metal Oxide Semiconductor Field Effect Transistor (MOSFET):	
- MOSFET Equivalent Circuit	
- MOSFET Mathematical Relations	

Operational Amplifiers	
Key Concepts	**Handbook Page References**
Inverting Operational Amplifier	
Non-Inverting Operational Amplifier	
Summing Amplifiers	
Common Mode Rejection Ratio (CMRR)	
Differential Amplifier	

Instrumentation	
Key Concepts	**Handbook Page References**
Measurement Uncertainty:	
- The Klime-McClintock Equation	

Sensors:	
- Temperature Transducers	
- Strain Transducers	
- Pressure Transducers	
Propagation of Error	

Power Electronics	
Key Concepts	**Handbook Page References**
Rectifiers:	
- Half-Wave Rectifiers	
- Full-Wave Rectifiers	
- DC Bus Voltage Ripple	
- Full-Wave Rectifier with Capacitor Filter	
- Single Phase Half Wave Rectifier	
- Single Phase Full Wave Rectifier	
- Three Phase Half Wave Rectifier	
- Three Phase Full Wave Rectifier	
Inverters	

NOTES/EQUATIONS SUMMARY

DIAGNOSTIC TEST CHECKLIST

Checklist Questions	Yes/No
Have you read and understood all the material in this chapter?	
Are you familiar with all the key concepts and formulas highlighted in the tables for this chapter?	
Have you filled in the tables with the correct NCEES® FE Reference Handbook page numbers for each listed concept and formula?	
Did you download and complete all the designated practice problems for this section?	
Did you review and understand the step-by-step solutions to the practice problems?	
Could you solve the practice problems without excessive reference to the solutions and within a reasonable time frame?	
Are you confident in applying the concepts and formulas from this chapter?	
Have you noted any areas of difficulty or topics that require further review before proceeding?	
Do you know how to quickly find information in the NCEES® FE Reference Handbook?	
Have you reviewed any errors or misunderstandings in this section's material to ensure clarity and understanding?	

CHAPTER 11

Power Systems

11.1 Power Generation and Distribution

Power generation involves converting various energy sources into electrical energy. These sources can be broadly categorized into fossil fuels, nuclear energy, and renewable energy.

- **Fossil Fuels:** These include coal, natural gas, and oil. They are the most common energy sources for power generation but contribute significantly to environmental pollution.

- **Nuclear Energy:** Uses nuclear reactions to generate heat, which is then converted into electrical energy. It provides a high energy output with low greenhouse gas emissions.

- **Renewable Energy:** Includes solar, wind, hydroelectric, and geothermal energy. These sources are sustainable and have a lower environmental impact compared to fossil fuels.

Conversion Methods

The process of converting primary energy sources into electrical energy varies depending on the type of energy source.

- **Thermal Conversion:** Used in fossil fuel and nuclear power plants where heat is used to produce steam that drives turbines connected to generators.
- **Photovoltaic Conversion:** In solar power plants, sunlight is directly converted into electricity using photovoltaic cells.
- **Mechanical Conversion:** Wind turbines and hydroelectric plants convert kinetic energy from wind or flowing water into mechanical energy, which is then converted into electrical energy by generators.

Example

- In a coal-fired power plant, coal is burned to heat water in a boiler, creating steam. The steam turns a turbine, which drives a generator to produce electricity.

Types of Power Plants

Fossil Fuel Power Plants

Fossil fuel power plants generate electricity by burning coal, natural gas, or oil to produce steam that drives turbines.

- **Coal-fired Plants:** Burn coal to generate steam. They are a major source of electricity but contribute to air pollution and greenhouse gas emissions.
- **Natural Gas Plants:** Use gas turbines or combined cycle systems. They are more efficient and cleaner than coal-fired plants.
- **Oil-fired Plants:** Burn oil to produce steam. They are less common due to higher fuel costs and environmental concerns.

Example:

- A natural gas plant might use a combined cycle system where gas turbines generate electricity, and the waste heat is used to produce additional electricity through steam turbines.

Nuclear Power Plants

Nuclear power plants use controlled nuclear reactions to generate heat, which is then used to produce steam and drive turbines.

- **Nuclear Fission:** Uranium or plutonium atoms are split in a reactor to release energy.
- **Nuclear Reactors:** Different types include pressurized water reactors (PWRs) and boiling water reactors (BWRs).

Example:

- A pressurized water reactor (PWR) keeps water under high pressure to prevent it from boiling, using the heat generated by nuclear fission to produce steam in a secondary loop.

Renewable Energy Power Plants

Renewable energy power plants utilize natural processes to generate electricity. They are environmentally friendly and sustainable.

- **Solar Power Plants:** Use photovoltaic cells to convert sunlight directly into electricity.
- **Wind Farms:** Use wind turbines to capture wind energy and convert it into electrical power.
- **Hydroelectric Plants:** Use the energy of falling or flowing water to drive turbines.

Example:

- A wind farm consists of multiple wind turbines that capture wind energy. The kinetic energy of the wind turns the turbine blades, generating electricity through an attached generator.

Power Distribution

Transmission Lines and Substations

Once electricity is generated, it needs to be transmitted over long distances to reach consumers.

- **Transmission Lines:** High-voltage lines that carry electricity from power plants to substations. They minimize energy loss over long distances.

- **Substations:** Facilities that step down the high voltage from transmission lines to lower voltages suitable for distribution to homes and businesses.

Example:

- Electricity generated at a power plant at 230 kV is transmitted over long distances and then stepped down to 11 kV at a substation for distribution to residential areas.

Distribution Networks

The distribution network carries electricity from substations to consumers. It includes local grids, transformers, and power lines.

- **Local Grids:** Networks of distribution lines that deliver electricity to homes, businesses, and industries.

- **Transformers:** Devices that adjust voltage levels to ensure safe and efficient power delivery.

Grid Infrastructure and Management

The power grid is a complex network that requires careful management to ensure reliability and stability.

- **Grid Management:** Involves monitoring and controlling the flow of electricity to balance supply and demand.
- **Smart Grids:** Advanced grids that use digital technology to improve reliability, efficiency, and sustainability.

Example

- Smart grids can automatically detect and respond to changes in electricity demand, integrating renewable energy sources more effectively.

Challenges and Innovations

Environmental Impact

Power generation, especially from fossil fuels, has significant environmental impacts, including air pollution and greenhouse gas emissions.

- **Air Pollution:** Emissions from coal and oil plants contribute to smog and respiratory issues.
- **Greenhouse Gases:** Fossil fuel combustion releases carbon dioxide, a major contributor to climate change.

Technological Advancements

Innovations in power generation and distribution aim to improve efficiency and reduce environmental impact.

- **Advanced Nuclear Reactors:** Safer and more efficient designs for nuclear power.

- **Improved Renewable Technologies:** Enhanced solar panels and wind turbines for greater energy capture and efficiency.
- **Example:**
- Advances in photovoltaic technology have led to more efficient solar panels that convert more sunlight into electricity, reducing the cost and increasing the viability of solar power.

Smart Grids

Smart grids integrate digital technology to enhance the efficiency, reliability, and sustainability of power distribution.

- **Benefits of Smart Grids:**
    - **Improved Reliability:** Real-time monitoring and automated fault detection.
    - **Enhanced Efficiency:** Better management of electricity flow and reduced energy losses.
    - **Sustainability:** Facilitates the integration of renewable energy sources.

Example

- A smart grid can detect a power outage and automatically reroute electricity to restore service quickly, minimizing downtime for consumers.

11.2 Transformers

Basic Principles and Functions

Transformers are electrical devices used to transfer electrical energy between two or more circuits through electromagnetic induction. They are essential in adjusting voltage levels to meet the requirements of different applications, ensuring efficient power distribution.

- **Principle of Operation:** Transformers operate on the principle of electromagnetic induction, where a varying current in the primary coil generates a magnetic field, inducing a voltage in the secondary coil.

- **Function:** The main functions of transformers include stepping up (increasing) or stepping down (decreasing) voltage levels, isolating circuits, and distributing power efficiently.

Example

- In a power plant, a step-up transformer increases the voltage for transmission over long distances, reducing energy loss. At the receiving end, a step-down transformer decreases the voltage to safe levels for use in homes and businesses.

Types of Transformers

Step-up and Step-down Transformers

- **Step-up Transformers:** Increase the voltage from the primary to the secondary coil. Used in power generation to transmit electricity over long distances.

- **Step-down Transformers:** Decrease the voltage from the primary to the secondary coil. Used to reduce voltage levels for safe use in residential and industrial settings.

Example

- A step-up transformer at a power station increases the voltage from 11 kV to 220 kV for efficient transmission. A step-down transformer near residential areas then decreases it from 220 kV to 11 kV or lower for household use.

Isolation Transformers

- **Purpose:** Provide electrical isolation between circuits without changing the voltage levels. Used for safety and noise reduction in sensitive equipment.
- **Function:** Isolate the primary and secondary windings, ensuring that electrical noise or faults do not transfer from one side to the other.

Example:

- Isolation transformers are used in medical equipment to protect patients and sensitive instruments from electrical faults.

Autotransformers

- **Design:** Use a single winding that acts as both the primary and secondary winding, with a tap that provides the desired voltage transformation.
- **Efficiency:** More efficient and compact than traditional transformers but provide less isolation.

Example:

- Autotransformers are used in voltage regulation applications, such as in variable voltage supplies for laboratory experiments.

Transformer Construction

Core Types

- **Laminated Core:** Made of thin, insulated layers of steel to reduce eddy current losses. Commonly used in power transformers.
- **Toroidal Core:** Doughnut-shaped core that provides high efficiency and low electromagnetic interference (EMI). Used in compact transformers.

Example:

- Laminated core transformers are used in power distribution due to their durability and efficiency, while toroidal core transformers are used in audio equipment for their low noise characteristics.

Windings

- **Primary Winding:** Connected to the input voltage source. Creates a magnetic field that induces voltage in the secondary winding.
- **Secondary Winding:** Connected to the output load. Receives induced voltage from the primary winding's magnetic field.
- **Taps:** Additional connections on the winding that allow for adjustment of voltage levels.

Example:

- A transformer with multiple taps on its secondary winding can provide different output voltages, making it versatile for various applications.

Applications of Transformers

Power Transmission and Distribution

Transformers play a critical role in the efficient transmission and distribution of electrical power from generation plants to consumers.

- **High-Voltage Transmission:** Step-up transformers increase voltage for transmission over long distances, reducing energy loss.
- **Distribution Networks:** Step-down transformers decrease voltage to safe levels for delivery to homes and businesses.

Example:

- In a typical power grid, electricity generated at 13.8 kV is stepped up to 230 kV for transmission. It is then stepped down to 33 kV at a substation and further reduced to 11 kV for distribution to residential areas.

Industrial and Residential Uses

Transformers are used in various industrial and residential applications to provide the required voltage levels for different equipment and appliances.

- **Industrial Applications:** Used in machinery, manufacturing equipment, and electric furnaces to provide appropriate voltage levels.

- **Residential Applications:** Found in household appliances, such as doorbell systems and power adapters, to convert voltage levels for safe operation.

Example

- In an industrial setting, a transformer steps down the high-voltage supply to a lower voltage suitable for operating heavy machinery. In homes, small transformers in power adapters convert the mains voltage to lower voltages for charging electronic devices.

11.3 Three-Phase Systems

Three-phase systems are a type of polyphase system used to generate, transmit, and distribute electrical power. They consist of three alternating currents (AC) of the same frequency and voltage amplitude but with a phase difference of 120 degrees between them.

Definition

A three-phase system has three sinusoidal voltages, each offset by 120 degrees, creating a constant power flow that is more efficient than single-phase systems.

Advantages

- **Efficiency:** Three-phase systems are more efficient in power generation and distribution.
- **Power Delivery:** They provide a constant power transfer, reducing pulsations and vibrations in machinery.
- **Economical:** They require less conductor material to transmit the same amount of power as single-phase systems.
- **Balanced Loads:** Three-phase systems can easily balance electrical loads, minimizing losses and improving stability.

Example

- In industrial settings, three-phase power is used to operate heavy machinery with smoother and more reliable power delivery than single-phase power.

Three-Phase Generation

Methods of Generation

Three-phase power is typically generated using three-phase alternators or generators, which convert mechanical energy into electrical energy.

- **Three-Phase Alternators:** Use rotating magnetic fields to induce three AC voltages in three separate windings. These voltages are 120 degrees out of phase with each other.
- **Mechanical Energy Sources:** The mechanical energy required to drive the alternators can come from various sources such as steam turbines, hydro turbines, wind turbines, and internal combustion engines.

Example

- A three-phase alternator in a power plant converts mechanical energy from a steam turbine into three-phase electrical power, which is then transmitted to the grid.

Three-Phase Connections

Star (Y) Connection

In a star (Y) connection, each of the three-phase windings is connected at a common point called the neutral, and the other ends are connected to the three-phase lines.

Characteristics

- **Neutral Point:** Provides a return path for unbalanced currents.
- **Phase Voltage (V_{phase}):** Voltage across each winding.
- **Line Voltage (V_{line}):** Voltage between any two lines.
- **Relationship:** Line voltage is √3 times the phase voltage.

$$V_{line} = \sqrt{3} \times V_{phase}$$

Example

In a star-connected system with a phase voltage of 230V, the line voltage is:

$$V_{line} = \sqrt{3} \times 230V \approx 400V$$

Delta (Δ) Connection

In a delta (Δ) connection, each of the three-phase windings is connected end-to-end, forming a closed loop.

Characteristics

- **No Neutral Point:** No common neutral point is required.
- **Phase Voltage and Line Voltage:** The phase voltage is equal to the line voltage.

- **Current Relationship:** Line current is √3 times the phase current.

$$I_{line} = \sqrt{3} \times I_{phase}$$

Example

In a delta-connected system with a phase current of 10A, the line current is:

$$I_{line} = \sqrt{3} \times 10A \approx 17.3A$$

Power Calculations in Three-Phase Systems

Line and Phase Voltages

Understanding the relationship between line and phase voltages and currents is crucial for power calculations in three-phase systems.

- **Star (Y) Connection:**
  - Line Voltage: $V_{line} = \sqrt{3} \times V_{phase}$
  - Line Current: $I_{line} = I_{phase}$
- **Delta (Δ) Connection:**
  - Line Voltage: $V_{line} = V_{phase}$
  - Line Current: $I_{line} = \sqrt{3} \times I_{phase}$

Power Formulas

The total power in a three-phase system can be calculated using the following formulas:

- **Active Power (P):** $P = \sqrt{3} \times V_{line} \times I_{line} \times \cos(\phi)$
  - Where ϕ is the phase angle between voltage and current.
- **Reactive Power (Q):** $Q = \sqrt{3} \times V_{line} \times I_{line} \times sin(\varphi)$
- **Apparent Power (S):** $S = \sqrt{3} \times V_{line} \times I_{line}$

Example

- For a star-connected system with a line voltage of 400V, line current of 10A, and a power factor (cos(φ)) of 0.8:

$$P = \sqrt{3} \times 400V \times 10A \times 0.8 \approx 5.54\,kW$$

$$Q = \sqrt{3} \times 400V \times 10A \times sin(cos^{-1}(0.8)) \approx 4.16\,kVAR$$

$$S = \sqrt{3} \times 400V \times 10A \approx 6.93\,kVA$$

Applications of Three-Phase Systems

Industrial Machinery

Three-phase systems are widely used in industrial settings due to their efficiency and reliability.

- **Advantages for Machinery**
  - Provide constant power delivery, reducing vibrations and mechanical stress.
  - Enable the use of three-phase motors, which are more efficient and have a higher power-to-weight ratio than single-phase motors.
- **Example:** A three-phase induction motor in a factory runs more smoothly and efficiently, reducing maintenance costs and downtime.

Power Distribution

Three-phase systems are essential for the transmission and distribution of electrical power.

- **Advantages for Distribution:**
  - Allow for the transmission of more power over long distances with less conductor material.
  - Provide a more balanced load, reducing losses and improving stability.

- **Example:** Electrical power generated at a power plant is transmitted using a three-phase system to ensure efficient delivery over long distances with minimal energy loss.

11.4 Power Factor Correction

Definition and Importance

Power factor is the ratio of real power (P) used to perform work to the apparent power (S) supplied to the circuit. It is a measure of how effectively electrical power is being used.

- **Formula:** $Power\ Factor = \frac{P}{S} = \cos(\phi)$
  - Where ϕ is the phase angle between the current and voltage.
- **Importance:**
  - High power factor indicates efficient utilization of electrical power.
  - Low power factor leads to increased power losses and reduced capacity of the electrical system.
- **Example:** In an industrial facility, if the power factor is 0.7, this means only 70% of the supplied power is being used for productive work, and the remaining 30% is wasted.

Causes of Low Power Factor

Inductive Loads

Inductive loads such as motors, transformers, and fluorescent lighting are common in industrial and commercial settings. These loads cause the current to lag behind the voltage, leading to a low power factor.

- **Characteristics**
  - Inductive loads consume reactive power (Q) in addition to real power (P).
  - This increases the total apparent power (S), lowering the power factor.
- **Example:** A motor in a factory consumes both real power to perform mechanical work and reactive power to maintain its magnetic field, resulting in a lagging power factor.

Reactive Power

Reactive power is the power that oscillates between the source and the load, without being converted into useful work. It is measured in volt-amperes reactive (VAR).

- **Impact**
  - High levels of reactive power lead to a lower power factor.
  - Reactive power does not contribute to the actual work done but increases the total power that must be supplied.
- **Example:** Fluorescent lights require reactive power to operate their ballasts, contributing to a lower power factor in the lighting system.

Methods of Power Factor Correction

Capacitors

Capacitors provide leading reactive power, which counteracts the lagging reactive power caused by inductive loads. Adding capacitors to the electrical system is a common method for improving power factor.

- **Function**
  - Capacitors store and release electrical energy, providing reactive power to the system.

- This reduces the phase difference between voltage and current, improving the power factor.

- **Example:** Installing capacitor banks in an industrial plant can offset the reactive power consumed by motors, thereby improving the overall power factor.

Synchronous Condensers

Synchronous condensers are essentially synchronous motors that operate without a mechanical load. They provide reactive power compensation and can adjust the power factor as needed.

- **Function**
  - Synchronous condensers can generate or absorb reactive power by adjusting their excitation.
  - They help maintain a desired power factor and stabilize the voltage in the system.
- **Example:** A synchronous condenser installed in a power distribution network can help balance reactive power demands and maintain a high power factor.

Active Power Factor Correction Devices

Active power factor correction (PFC) devices use power electronics to improve the power factor dynamically.

- **Function**
  - These devices continuously monitor the power factor and inject the necessary reactive power to correct it.
  - They are more precise and efficient compared to passive methods like capacitors.

- **Example:** An active PFC device in a data center adjusts the power factor in real time, ensuring efficient power usage and reducing energy costs.

Benefits of Power Factor Correction

Efficiency Improvements

Improving the power factor reduces the total current flowing through the system, which decreases power losses in the conductors and transformers.

- **Benefits**
    - Reduced energy losses result in lower operating temperatures and extended equipment life.
    - Enhanced capacity of the electrical system, allowing more loads to be connected without upgrading the infrastructure.
- **Example:** A manufacturing facility with improved power factor experiences less heat generation in its electrical cables, reducing maintenance costs and improving reliability.

Cost Savings

Higher power factors can lead to significant cost savings, both in terms of reduced energy bills and lower demand charges from utility companies.

- **Benefits**
    - Utility companies often charge penalties for low power factor, so improving it can reduce these charges.
    - Increased efficiency means lower electricity consumption, translating to lower energy bills.
- **Example:** By implementing power factor correction, a commercial building can lower its electricity costs and avoid penalties, leading to substantial financial savings over time.

ADDITIONAL ONLINE RESOURCES

- **1500+ PRACTICE PROBLEMS [4 PRACTICE EXAMS INCLUDED]**

Enhance your exam preparation with over 1000+ practice problems, complete with step-by-step solutions. Two weeks after your initial download, you will receive a weekly full-length practice exam with step-by-step solutions for four weeks, totaling over 1500 practice problems.

SCAN HERE

Having trouble accessing the resources? Email us at **info@learnovapublishing.com** — we reply within 24 hours, 7 days a week

- **VIDEO RESOURCE RECOMMENDATIONS**

Enhance your learning with our recommended video resources for a deeper understanding of the concepts. When using these resources, please check the current exam specifications.

An abundance of high-quality video resources is available for preparing for this section. For a comprehensive overview, we recommend the following:

YouTube Playlist: *FE Electrical and Computer: Review (Effective from July 2020) by Amit Roy*
Search for this playlist on YouTube. It covers almost all the exam topics, providing detailed explanations and condensing the material effectively.

Online Review Course: For a more organized approach, consider the comprehensive, free, video-based online review course called *"Electrical FE Review"* by Zach Stone, available at electricalfereview.com

Power Theory	
Key Concepts	**Handbook Page References**
Apparent, Real and Reactive Power	
Single-Phase Power	
Three-Phase Power:	
- Delta Arrangements	
- Wye Arrangements	
- Conversion between Delta and Wye	
Power Factor:	
- Waveform - Voltage & Current	
- Phasor - Voltage & Current	
- Power Triangle	

Transmission and Distribution	
Key Concepts	**Handbook Page References**
Transmission Line Parameters	
- Series Resistance	
- Series Inductance	
- Shunt Capacitance	
- Shunt Conductance	
Voltage Regulation	
Voltage Drop	
Losses and Efficiency	

Transformers	
Key Concepts	**Handbook Page References**
Transformers Types	
Transformer Construction	
Ideal Transformer	
Turns Ratio	
Impedance Transformation	

Motors and Generators	
Key Concepts	**Handbook Page References**
Construction and Types:	
- Synchronous Motors and Generators	
- Induction Motors and Generators	
- Servomotors	
Synchronous Speed	
Slip	
Power vs Torque	
Torque-Speed Characteristics	

NOTES/EQUATIONS SUMMARY

DIAGNOSTIC TEST CHECKLIST

Checklist Questions	Yes/No
Have you read and understood all the material in this chapter?	
Are you familiar with all the key concepts and formulas highlighted in the tables for this chapter?	
Have you filled in the tables with the correct NCEES® FE Reference Handbook page numbers for each listed concept and formula?	
Did you download and complete all the designated practice problems for this section?	
Did you review and understand the step-by-step solutions to the practice problems?	
Could you solve the practice problems without excessive reference to the solutions and within a reasonable time frame?	
Are you confident in applying the concepts and formulas from this chapter?	
Have you noted any areas of difficulty or topics that require further review before proceeding?	
Do you know how to quickly find information in the NCEES® FE Reference Handbook?	
Have you reviewed any errors or misunderstandings in this section's material to ensure clarity and understanding?	

CHAPTER 12

Electromagnetics

Electromagnetics is a fundamental area of electrical engineering that deals with the study of electric and magnetic fields and their interactions. It encompasses a wide range of phenomena and principles that are crucial for understanding how various electrical devices and systems operate. This chapter aims to provide a comprehensive overview of key topics in electromagnetics, including electrostatics, magnetostatics, electromagnetic induction, and transmission lines.

12.1 Electrostatics

Electrostatics is the branch of physics that studies electric charges at rest. It involves understanding the forces, fields, and potentials associated with static electric charges. Electrostatics is fundamental to many applications in electrical engineering, such as designing capacitors, analyzing electric fields, and ensuring proper insulation in electrical systems.

Coulomb's Law

Coulomb's Law quantifies the force between two point charges. It states that the electric force between two charges is directly proportional to the product of the charges but inversely proportional to the square of the distance between them. The law is expressed as: $F = k_e \frac{q_1 q_2}{r2}$

- F: Force between the charges (Newtons)
- k_e: Coulomb's constant ($8.99 \times 10^9\, N\, m^2/C^2$)
- q_1, q_2: Magnitudes of the charges (Coulombs)
- r: Distance between the charges (meters)

Example

If two charges of 1 μC each are separated by a distance of 1 meter, the force between them is: $F = 8.99 \times 10^9 \frac{(1 \times 10^{-6})^2}{1^2} = 8.99\, N$

This example illustrates how the force increases with the magnitude of the charges and decreases with the square of the distance between them.

Electric Field and Electric Potential

An electric field (E) is a region around a charged particle where a force would be exerted on other charges. It is defined as the force per unit charge and is expressed as: $E = \frac{F}{q} = k_e \frac{q}{r^2}$

- E: Electric field (Newtons per Coulomb or Volts per meter)
- F: Force experienced by a test charge (Newtons)
- q: Test charge (Coulombs)

Example

The electric field 2 meters away from a charge of $5\, \mu C$ is:

$$E = 8.99 \times 10^9 \frac{(5 \times 10^{-6})}{2^2} = 11.24 \times 10^3\, N/C$$

This demonstrates how the electric field strength decreases with the square of the distance from the charge.

Electric potential (V) is the work done per unit charge in bringing a charge from infinity to a point in space. It is given by: $V = k_e \frac{q}{r}$

- **V**: Electric potential (Volts)
- **q**: Charge (Coulombs)
- **r**: Distance from the charge (meters)

Capacitance and Capacitors

Capacitance (C) is the ability of a system to store charge per unit voltage. Capacitors are devices that store electrical energy in an electric field. The capacitance of a capacitor is defined as: $C = \frac{Q}{V}$

- **C**: Capacitance (Farads)
- **Q**: Charge stored (Coulombs)
- **V**: Voltage across the capacitor (Volts)

There are various types of capacitors, including parallel plate capacitors, cylindrical capacitors, and spherical capacitors. The most common type, the parallel plate capacitor, has a capacitance given by: $C = \epsilon_0 \epsilon_r \frac{A}{d}$

- ϵ_0 : Permittivity of free space ($8.85 \times 10^{-12}\ F/m$)
- ϵ_r: Relative permittivity (dielectric constant)
- **A**: Area of the plates (square meters)
- **d**: Distance between the plates (meters)

Example

A parallel plate capacitor with plates of area 1 square meter separated by 1 millimeter and a dielectric constant of 5 has a capacitance of:

$$C = 8.85 \times 10^{-12} \times 5 \times \frac{1}{1 \times 10^{-3}} = 44.25\,pF$$

This shows how capacitance increases with the area of the plates and the dielectric constant and decreases with the distance between the plates.

Real-world Examples

Capacitors in Circuits

Capacitors are widely used in electronic circuits for energy storage, filtering, and timing applications. In power supplies, capacitors smooth out fluctuations in voltage, providing a stable output.

Electrostatic Precipitators

In industrial applications, electrostatic precipitators remove particles from exhaust gases. They use high-voltage electrodes to create an electric field that charges the particles, which are then attracted to and collected on oppositely charged plates.

- **Example:** In a power supply circuit, capacitors filter out AC ripple from the DC output, ensuring a steady voltage to sensitive electronic components.

Insulation

Understanding electrostatics is crucial for designing insulation systems in high-voltage equipment. Proper insulation prevents unwanted discharges and protects equipment from damage.

- **Example:** In high-voltage transmission lines, insulators prevent the current from leaking into the supporting structures, ensuring efficient power delivery and safety.

12.2 Magnetostatics

Magnetostatics is the study of magnetic fields in systems where the currents are steady and not changing with time. It deals with the forces and fields associated with magnetic materials and currents. Understanding magnetostatics is crucial for designing and analyzing devices like transformers, inductors, and motors.

Biot-Savart Law

The Biot-Savart Law describes the magnetic field generated by a steady current. It states that the magnetic field (B) at a point is directly proportional to the current (I) and inversely proportional to the square of the distance (r) from the current element.

Formula

$$B = \frac{\mu_0}{4\pi}\frac{I \cdot dl \times \hat{r}}{r^2}$$

- Where B is the magnetic field, μ_0 is the permeability of free space ($4\pi \times 10^{-7} Tm/A$), I is the current, dl is the length element of the current-carrying conductor, and $\hat{r}$ is the unit vector from the current element to the point of interest.

Example:

- The magnetic field at the center of a circular loop of radius R carrying current I is: $B = \frac{\mu_0 I}{2R}$

This formula shows how the magnetic field strength decreases with the distance from the current-carrying conductor.

Consider a circular loop of wire with a radius of 0.1 meters carrying a current of 5 Amperes. The magnetic field at the center of the loop can be calculated using the formula: $B = \frac{\mu_0 I}{2R}$

Substituting the values: $B = \frac{4\pi \times 10^{-7} \times 5}{(2 \times 0.1)+} = 3.14 \times 10^{-5}$

This example illustrates how the magnetic field at the center of the loop depends on the current and the radius of the loop.

Ampere's Law

Ampere's Law relates the magnetic field along a closed loop to the total current passing through the loop. It simplifies the calculation of magnetic fields in symmetrical situations and is fundamental in understanding magnetic circuits.

Formula

$$\oint B \cdot dl = \mu_0 I_{enc}$$

- Where $\oint B \cdot dl$ is the line integral of the magnetic field around a closed path, and I_{enc} is the total current enclosed by the path.

Example

- For a long straight wire carrying current I, the magnetic field at a distance r from the wire is: $B = \frac{\mu_0 I}{2\pi r}$

Consider a long straight wire carrying a current of 10 Amperes. To find the magnetic field at a distance of 0.05 meters from the wire, we use the formula: $B = \frac{\mu_0 I}{2\pi r}$

Substituting the values: $B = \frac{4\pi \times 10^{-7} \times 10}{2\pi \times 0.05} = 4 \times 10^{-5}\,T$

This example shows how the magnetic field around a straight wire decreases with distance.

Magnetic Materials and Magnetic Circuits

Magnetic materials can be classified into diamagnetic, paramagnetic, and ferromagnetic based on their response to an external magnetic field. Magnetic circuits are analogous to electric circuits but involve magnetic fields and flux.

Magnetic Materials

- **Diamagnetic:** Weakly repelled by magnetic fields (e.g., copper).

- **Paramagnetic:** Weakly attracted by magnetic fields (e.g., aluminum).
- **Ferromagnetic:** Strongly attracted by magnetic fields (e.g., iron, cobalt, nickel).

Magnetic Circuits

- **Magnetic Flux (Φ):** The total magnetic field passing through a surface.
- **Magnetomotive Force (MMF):** The magnetic potential difference driving the flux, analogous to voltage in electric circuits.
- **Reluctance (R_m):** Opposition to magnetic flux, analogous to resistance in electric circuits.

Example

- An iron core inductor has a high magnetic permeability, allowing it to concentrate and enhance the magnetic flux within its coils. The MMF drives the magnetic flux through the core, and the reluctance determines how easily the flux can flow.
- Consider an iron core with a permeability (μ) of 2000 × μ_0, a cross-sectional area (A) of 0.01 m^2, and a length (l) of 0.1 meters. The magnetic flux (Φ) generated by an MMF of 1000 Ampere-turns can be calculated using: $\Phi = \frac{MMF}{R_m}$

Where: $Rm = \frac{l}{\mu A}$

$$Rm = \frac{0.1}{2000 \times 4\pi \times 10^{-7} \times 0.01} = 398 \text{ A-turns/Wb}$$

$$\Phi = \frac{1000}{398} = 2.51\, Weber$$

This example shows how to calculate the magnetic flux in a magnetic circuit using the given parameters.

Real-world Examples

Transformers

Transformers use magnetic circuits to transfer electrical energy between circuits through electromagnetic induction. The core of a transformer is made of ferromagnetic material to efficiently direct the magnetic flux.

Example

- A transformer in a power supply steps down high voltage from the power lines to a lower voltage suitable for household use. The magnetic core ensures efficient energy transfer with minimal losses.

- Consider a transformer with a primary winding of 1000 turns and a secondary winding of 100 turns. If the primary voltage is 230V, the secondary voltage can be calculated using the turns ratio: $\frac{Vp}{Vs} = \frac{Np}{Ns}$

$$\frac{Vp}{Vs} = \frac{Np}{Ns}$$

$$Vs = \frac{Vp \times Ns}{Np} = \frac{230 \times 100}{1000} = 23V$$

This example demonstrates how transformers adjust voltage levels for different applications.

Electric Motors

Electric motors convert electrical energy into mechanical work using magnetic fields. They rely on the interaction between magnetic fields and current-carrying conductors to produce torque.

Example

- In a DC motor, current flowing through the armature windings generates a magnetic field that interacts with the magnetic field of the stator, producing rotational motion.

- Consider a simple DC motor with a 12V supply and an armature resistance of 1 ohm. The current through the armature can be calculated as:

$$I = \frac{V}{R} = \frac{12}{1} = 12A$$

If the motor's torque constant is 0.1 Nm/A, the torque produced is:

$$T = K_t \times I = 0.1 \times 12 = 1.2\,Nm$$

This example shows how to calculate the torque in a DC motor.

Magnetic Storage

Magnetic materials are used in data storage devices such as hard drives, where data is stored by magnetizing tiny regions of a ferromagnetic material.

Example

- In a hard drive, data is written by changing the magnetization of small regions on the disk's surface. A read/write head detects these changes, allowing data to be read back.
- Consider a hard drive with a storage capacity of 1 TB. The drive uses magnetic domains of 100 nm in size. The number of magnetic domains can be estimated by dividing the total storage capacity by the size of each domain:

$$Number\ of\ domains = \frac{1TB}{100nm} \approx 10^{12}$$

This example illustrates the high density of data storage achieved with magnetic materials.

Magnetic Resonance Imaging (MRI)

MRI machines use strong magnetic fields and radio waves to create detailed images of the inside of the body.

Example

- An MRI machine generates a powerful magnetic field that aligns the nuclei of certain atoms in the body. Radio waves are then used to alter this alignment, and the machine measures the resulting signals to create images.

- Consider an MRI machine with a magnetic field strength of 3 Tesla. The frequency of the radio waves used to excite hydrogen nuclei can be calculated using the Larmor frequency: $f = \gamma B$

Where γ is the gyromagnetic ratio for hydrogen ($42.58 \times 10^6\ Hz/T$)

$$f = 42.58 \times 10^6 \times 3 = 127.74 \times 10^6\ Hz = 127.74\ MHz$$

This example demonstrates how the frequency of the radio waves used in MRI is determined by the strength of the magnetic field and the properties of the nuclei being imaged.

12.3 Electromagnetic Induction

Electromagnetic induction is the process by which a changing magnetic field induces an electromotive force (EMF) in a conductor. This principle is fundamental to many technologies, including transformers, electric generators, and inductors. Understanding electromagnetic induction helps engineers design systems that convert mechanical energy to electrical energy and vice versa.

Faraday's Law of Induction

Faraday's Law of Induction states that the induced EMF in a closed circuit is directly proportional to the rate of change of magnetic flux through the circuit. The law is expressed as: $\epsilon = -\frac{d\Phi_B}{dt}$

- ϵ: Induced EMF (volts)

- Φ_B: Magnetic flux (webers)

- $d\Phi_B/dt$: Rate of change of magnetic flux

Explanation

- **Magnetic Flux (Φ):** The product of the magnetic field (B) and the area (A) through which the field lines pass, $\Phi_B = B \cdot A \cdot cos(\theta)$ where θ is the angle between the magnetic field and the normal to the surface. The negative sign indicates the direction of the induced EMF, as given by Lenz's Law.

- Consider a loop of wire with an area of 0.1 m^2 placed in a magnetic field that increases uniformly from 0 to 2 T in 1 second. The induced EMF can be calculated using Faraday's Law:

$$\epsilon = -\frac{d\Phi_B}{dt} = -\frac{d(B \cdot A)}{dt} = -\frac{2\,T \cdot 0.1\,m^2 - 0\,T \cdot 0.1\,m^2}{1s} = -0.2\,V$$

  This example shows how the EMF is induced by a changing magnetic field.

Lenz's Law

Lenz's Law states that the direction of the induced current in a conductor is such that it opposes the change in magnetic flux that produced it. This is a consequence of the conservation of energy and is reflected in the negative sign in Faraday's Law.

Explanation

- When a magnetic field changes, the induced EMF generates a current that creates its own magnetic field. This induced magnetic field opposes the change in the original magnetic field.

- If the magnetic field through a loop is increasing, the induced current will flow in a direction that creates a magnetic field opposing the increase. Conversely, if the magnetic field is decreasing, the induced current will create a magnetic field that opposes the decrease.

Inductors and Inductance

Inductance is the property of a conductor by which a change in current induces an EMF in both the conductor itself (self-inductance) and in nearby conductors (mutual inductance). Inductors are components that utilize this property to store energy in a magnetic field.

- **Self-Inductance (L):** The EMF induced in a conductor due to a change in its own current. Given by: $\epsilon_L = -L\frac{dI}{dt}$
  - **L:** Inductance (Henries)
  - **dI/dt:** Rate of change of current
- **Mutual Inductance (M):** The EMF induced in one conductor due to a change in current in another nearby conductor. Given by: $\epsilon_M = -M\frac{dI_2}{dt}$
  - **M:** Mutual inductance (Henries)
  - dI_2/dt: Rate of change of current in the second conductor

Example

- Consider an inductor with an inductance of 2 H and a current that changes at a rate of 3 A/s. The induced EMF can be calculated as:

$$EL = -L\frac{dI}{dt} = -2 \times 3 = -6\,V$$

This example shows how inductors generate an EMF in response to changes in current.

Real-world Examples

Transformers

Transformers rely on electromagnetic induction to transfer electrical energy between circuits. A changing current in the primary coil generates a changing magnetic field, which induces an EMF in the secondary coil.

Example

- A transformer with 1000 turns in the primary coil and 100 turns in the secondary coil. If the primary voltage is 230V, the secondary voltage can be calculated using the turns ratio: $\frac{Vp}{Vs} = \frac{N_p}{N_s}$

$$V_s = \frac{Vp \times Ns}{Np} = \frac{230 \times 100}{1000} = 23V$$

This shows how transformers adjust voltage levels for different applications.

Electric Generators

Electric generators convert mechanical energy into electrical energy using electromagnetic induction. As the generator's coil rotates in a magnetic field, a varying magnetic flux induces an EMF in the coil.

Example

- In a simple AC generator, a coil of wire rotates in a magnetic field at a constant speed. If the coil has 200 turns and rotates at 50 Hz, with a peak magnetic flux of 0.1 Wb, the maximum EMF induced is:

$$\epsilon = N \cdot \frac{d\Phi}{dt} = 200 \times 2\pi \times 50 \times 0.1 = 6283\,V$$

This example illustrates how generators use electromagnetic induction to produce electricity.

Induction Cooktops

Induction cooktops use electromagnetic induction to heat cookware. An alternating current in the cooktop's coil generates a changing magnetic field, inducing currents in the cookware and producing heat.

Example

- In an induction cooktop, a coil generates a high-frequency alternating current, creating a rapidly changing magnetic field. This field induces eddy currents in the ferromagnetic cookware, which resists the flow of electricity and generates heat, cooking the food.

12.4 Transmission Lines

Transmission lines are specialized cables or other structures designed to carry alternating current (AC) and radio frequency (RF) signals from one point to another with minimal loss. They are essential for power distribution, telecommunications, and various other applications in electrical and electronic engineering.

Example

- Overhead power lines carry high-voltage electricity from power plants to substations, where it is stepped down for distribution to homes and businesses.

Transmission Line Parameters

Transmission lines are characterized by several key parameters that determine their performance:

- **Resistance (R):** Opposition to current flow, causing power loss as heat.
- **Inductance (L):** Property that causes opposition to changes in current, storing energy in a magnetic field.

- **Capacitance (C):** Ability to store electric charge, causing opposition to changes in voltage.
- **Conductance (G):** Measure of a transmission line's ability to conduct electric current, accounting for dielectric losses.

Explanation

- The primary parameters (R, L, C, G) define the transmission line's behavior and are used to model and analyze its performance. They affect the line's impedance, signal attenuation, and phase shift.

Example

- For a typical coaxial cable, the inductance per unit length might be 250 nH/m, and the capacitance per unit length might be 100 pF/m. These values help determine the characteristic impedance and signal propagation speed.

Wave Propagation and Reflection

Electromagnetic waves propagate along transmission lines, and their behavior is influenced by the line's parameters. Understanding wave propagation and reflection is crucial for ensuring efficient signal transmission.

- **Propagation:** Electromagnetic waves travel along the transmission line at a speed determined by the inductance and capacitance.
- **Reflection:** Occurs when there is a mismatch in impedance at the end of the transmission line, causing part of the signal to be reflected.

Explanation

- The velocity of wave propagation (v) is given by: $v = \frac{1}{\sqrt{LC}}$
- Reflection is characterized by the reflection coefficient (Γ): $\Gamma = \frac{Z_L - Z_0}{Z_L + Z_0}$

  Where Z_L is the load impedance and Z_0 is the characteristic impedance of the line.

Example

- If a transmission line with a characteristic impedance of 50 ohms is terminated with a load impedance of 100 ohms, the reflection coefficient is:

$$\Gamma = \frac{100 - 50}{100 + 50} = \frac{50}{150} = 0.33$$

This indicates that 33% of the signal will be reflected back toward the source.

Impedance Matching

Impedance matching is the process of making the load impedance equal to the source or transmission line impedance to maximize power transfer and minimize reflections.

- **Importance:** Prevents signal loss and distortion, ensuring efficient power transfer and high-quality signal transmission.

- **Methods:** Include using transformers, matching networks, and stubs to adjust impedance.

Explanation

- Impedance matching devices, such as quarter-wave transformers, are used to match impedances. A quarter-wave transformer has a characteristic impedance (Z_t) given by: $Z_t = \sqrt{Z_s Z_L}$

  Where Z_s is the source impedance and Z_L is the load impedance.

Example

- To match a 75-ohm source to a 300-ohm load using a quarter-wave transformer, the transformer's characteristic impedance should be:

$$Z_t = \sqrt{75 \times 300} = 150\, ohms$$

Real-world Examples

Power Transmission

High-voltage transmission lines transport electricity over long distances with minimal losses. The high voltage reduces the current, minimizing resistive losses in the conductors.

A 500 kV transmission line carrying 1000 MW of power over 200 km. By using high voltage, the current is reduced to 2000 A, significantly lowering resistive losses compared to lower voltage transmission.

Telecommunications

Coaxial and twisted-pair cables are used to transmit data and voice signals with high fidelity. Proper impedance matching and shielding minimize signal degradation.

In a coaxial cable TV system, the cables are designed with a characteristic impedance of 75 ohms. Proper connectors and impedance matching ensure clear signal transmission without reflections or signal loss.

RF Transmission

Waveguides and RF cables are used in wireless communication systems to transmit radio frequency signals with minimal loss. Impedance matching is critical for efficient signal transfer.

An RF transmitter operating at 2.4 GHz uses a coaxial cable with a characteristic impedance of 50 ohms. Matching the impedance of the antenna to the cable ensures maximum signal power is radiated, enhancing communication range and quality.

ADDITIONAL ONLINE RESOURCES

- **1500+ PRACTICE PROBLEMS [4 PRACTICE EXAMS INCLUDED]**

Enhance your exam preparation with over 1000+ practice problems, complete with step-by-step solutions. Two weeks after your initial download, you will receive a weekly full-length practice exam with step-by-step solutions for four weeks, totaling over 1500 practice problems.

SCAN HERE

Having trouble accessing the resources? Email us at **info@learnovapublishing.com** — we reply within 24 hours, 7 days a week

- **VIDEO RESOURCE RECOMMENDATIONS**

Enhance your learning with our recommended video resources for a deeper understanding of the concepts. When using these resources, please check the current exam specifications.

An abundance of high-quality video resources is available for preparing for this section. For a comprehensive overview, we recommend the following:

YouTube Playlist: *FE Electrical and Computer: Review (Effective from July 2020) by Amit Roy*

Search for this playlist on YouTube. It covers almost all the exam topics, providing detailed explanations and condensing the material effectively.

Online Review Course: For a more organized approach, consider the comprehensive, free, video-based online review course called *"Electrical FE Review"* by Zach Stone, available at electricalfereview.com

Electrostatics and Magnetostatics	
Key Concepts	**Handbook Page References**
Coulomb's Law:	
- Point Charge	
- Line Charge	
- Sheet Charge	
Forces on Charged Particles	
Electric Field Due to a Point Charge	
Gauss' Law for Electricity	
Work and Energy in Electric Fields	
Current Induced in Magnetic Field:	
- Magnetic Field Intensity Along a Wire (Vector Form, Cylindrical Coordinates)	
Magnetic Force on Current-Carrying Conductor:	
- Force on a Moving Charge in a Magnetic Field	
Energy Stored in a Magnetic Field	

Electrodynamics	
Key Concepts	**Handbook Page References**
Maxwell's Equations:	
- Gauss' Law for Magnetism	

- Faraday's Law of Induction	
- Ampere's Law	
Right Hand Rule:	
- Moving Charge	
- Current Moving through a Wire	
- Current through Coil & Solenoid	
Sinusoidal Motion:	
- Wavelength of a Sinusoidal Signal	

Transmission Lines	
Key Concepts	**Handbook Page References**
Transmission Line Parameters:	
- Propagation Constant	
- Characteristic Impedance	
- Transmission Line Impedance	
- Transmission Line Voltage	
- Transmission Line Current	
Reflection Coefficient	
Standing Wave Ratio	
Wave Equation	

NOTES/EQUATIONS SUMMARY

DIAGNOSTIC TEST CHECKLIST

Checklist Questions	Yes/No
Have you read and understood all the material in this chapter?	
Are you familiar with all the key concepts and formulas highlighted in the tables for this chapter?	
Have you filled in the tables with the correct NCEES® FE Reference Handbook page numbers for each listed concept and formula?	
Did you download and complete all the designated practice problems for this section?	
Did you review and understand the step-by-step solutions to the practice problems?	
Could you solve the practice problems without excessive reference to the solutions and within a reasonable time frame?	
Are you confident in applying the concepts and formulas from this chapter?	
Have you noted any areas of difficulty or topics that require further review before proceeding?	
Do you know how to quickly find information in the NCEES® FE Reference Handbook?	
Have you reviewed any errors or misunderstandings in this section's material to ensure clarity and understanding?	

CHAPTER 13

Control Systems

Control systems are essential in engineering for managing and regulating the behavior of dynamic systems. They are used in a wide range of applications, from simple household appliances to complex industrial processes and advanced robotics. A control system ensures that a system's output follows a desired input or setpoint by adjusting its behavior based on feedback.

Understanding control systems involves learning how to model physical systems, analyze their stability, and design appropriate feedback mechanisms and controllers. This knowledge is crucial for developing systems that are both efficient and reliable.

13.1 System Modeling

System modeling is the process of creating a mathematical representation of a physical system. This representation helps engineers analyze, understand, and predict the system's behavior under various conditions. Modeling is a fundamental step in the design and control of engineering systems, providing insights into how systems respond to inputs and how they can be controlled.

Mathematical Models

Mathematical models use equations to describe the dynamics of a system. These models can be linear or nonlinear, continuous or discrete, depending on the nature of the system.

- **Linear Models:** Simplify analysis and control design but may not capture all real-world behaviors.
- **Nonlinear Models:** More accurate but complex to analyze and control.
- **Continuous Models:** Represent systems where changes occur smoothly over time. Often described by differential equations.

Example

- The motion of a mass-spring-damper system can be modeled using Newton's second law. The equation of motion is: $m\ddot{x}(t) + b\dot{x}(t) + kx(t) = F(t)$
  - **m:** Mass
  - **b:** Damping coefficient
  - **k:** Spring constant
  - **x:** Displacement
  - **F(t):** External force
- **Discrete Models:** Represent systems where changes occur at specific intervals. Often described by differential equations.

Transfer Functions

A transfer function is a mathematical representation that relates the output of a system to its input in the Laplace domain. It is particularly useful for analyzing linear time-invariant (LTI) systems.

- **Definition:** The ratio of the Laplace transform of the output to the Laplace transform of the input, assuming zero initial conditions.
- **Formula:** $G(s) = \frac{Y(s)}{U(s)}$
  - **G(s):** Transfer function
  - **Y(s):** Output in the Laplace domain
  - **U(s):** Input in the Laplace domain

Example

- For the mass-spring-damper system, taking the Laplace transform and solving for the transfer function: $G(s) = \frac{X(s)}{F(s)} = \frac{1}{ms^2 + bs + k}$

  This transfer function describes how the displacement X(s) responds to an external force F(s).

State-Space Representation

State-space representation provides a comprehensive way to model dynamic systems, especially those with multiple inputs and outputs. It uses a set of first-order differential equations to describe the system's state variables.

- **State Variables:** Variables that represent the system's state at any given time.
- **State Equations:** A set of first-order differential equations that describe the dynamics of the state variables.

Form

$$\dot{x}(t) = Ax(t) + Bu(t)$$

$$y(t) = Cx(t) + Du(t)$$

- **x(t):** State vector
- **u(t):** Input vector

- **y(t):** Output vector
- **A:** State matrix
- **B:** Input matrix
- **C:** Output matrix
- **D:** Feedforward matrix

Example

- For the mass-spring-damper system, choosing $x_1 = x \; and \; x_2 = x^{\cdot}$ as state variables, the state-space representation is:

$$\begin{pmatrix} x_1^{\cdot} \\ x_2^{\cdot} \end{pmatrix} = \begin{pmatrix} 0 & 1 \\ -\frac{k}{m} & -\frac{b}{m} \end{pmatrix} \begin{pmatrix} x_1 \\ x_2 \end{pmatrix} + \begin{pmatrix} 0 \\ \frac{1}{m} \end{pmatrix} F(t)$$

$$y(t) = (1 \;\; 0) \left(\frac{x_1}{x_2}\right)$$

Real-world Examples

Automobile Suspension System

- Modeled as a mass-spring-damper system to design suspensions that provide comfort and stability.
- **Model**

$$mx^{\cdot\cdot} + bx^{\cdot} + kx = F(t)$$

- **Application:** Analyzing how the suspension responds to road irregularities.

Electric Circuit

- Modeled using Kirchhoff's laws to design and analyze circuits.
- **Model**

$$L\left(\frac{dI}{dt}\right) + RI + \frac{1}{C}\int I \, dt = V(t)$$

- **Application:** Determining the current and voltage behavior in an RLC circuit.

Thermal System:

- Modeled to manage heat transfer in systems like HVAC.
- **Model**

$$C\frac{dT}{dt} + \frac{1}{R}(T - T_{env}) = Q(t)$$

- **Application:** Controlling temperature in a building by analyzing heat flow and thermal properties.

13.2 Stability Analysis

Stability in control systems refers to the ability of a system to return to equilibrium after a disturbance. A stable system will respond to an input or disturbance without oscillating uncontrollably or diverging. Stability analysis is crucial in designing systems that perform reliably and predictably.

Methods of Stability Analysis

Routh-Hurwitz Criterion

The Routh-Hurwitz criterion is a mathematical technique used to determine the stability of a linear time-invariant (LTI) system by examining the characteristic equation of the system.

- **Steps**
    - Write the characteristic equation in the form $a_0 s^n + a_1 s^{n-1} + \cdots + a_n = 0$
    - Construct the Routh array using the coefficients of the characteristic equation.
    - Check the first column of the Routh array. If all elements in the first column are positive, the system is stable.

- **Example**

For the characteristic equation $s^3 + 2s^2 + 3s + 4 = 0$

$$\begin{array}{c|cc} s^3 & 1 & 3 \\ s^2 & 2 & 4 \\ s^1 & \frac{2\cdot3-1\cdot4}{2}=1 & 0 \\ s^0 & 4 & \end{array}$$

Since all elements in the first column are positive, the system is stable.

Nyquist Criterion

The Nyquist criterion assesses the stability of a closed-loop system by analyzing the frequency response of its open-loop transfer function. It involves plotting the Nyquist plot and checking for encirclements of the critical point (-1, 0) in the complex plane.

- **Steps**
  - Plot the Nyquist diagram of the open-loop transfer function.
  - Count the encirclements of the point (-1, 0).
  - Apply the Nyquist criterion to determine the stability based on the number of encirclements.

- **Example**

Consider a system with an open-loop transfer function

$$G(s)H(s) = \frac{10}{s(s+2)(s+5)}$$

Plot the Nyquist diagram and check for encirclements around the critical point (-1, 0). If the plot encircles (-1, 0) once clockwise, the system has one unstable pole, indicating instability.

Root Locus

The root locus method provides a graphical way to study how the roots of the characteristic equation change as a system parameter varies. It helps determine system stability and design controllers.

- **Steps**
  - Plot the root locus of the characteristic equation.
  - Analyze the location of poles and zeros to determine stability.
  - Adjust system parameters to place the poles in desired locations for stability.
- **Example**

  For a transfer function $G(s) = \frac{K}{(s(s+2)(s+4))}$, the root locus shows how the poles move as the gain K changes. By plotting the root locus, we can determine the range of K values that keep the poles in the left half-plane, ensuring stability. For instance, if K is between 0 and 10, the system remains stable.

Bode Plots

Bode plots are used to analyze the frequency response of a system. They consist of two plots: magnitude plot (gain vs. frequency) and phase plot (phase angle vs. frequency).

- **Components**
  - **Magnitude Plot:** Shows the gain of the system as a function of frequency.
  - **Phase Plot:** Shows the phase shift introduced by the system as a function of frequency.
  - **Gain Margin and Phase Margin:** Indicators of system stability derived from Bode plots.

- **Example**

For a system with a transfer function $G(s) = \frac{10}{s(s+1)(s+10)}$, plot the Bode plot to determine the gain and phase margins. The gain margin is the amount of gain increase required to make the system unstable, and the phase margin is the additional phase lag required to make the system unstable. If the gain margin is 20 dB and the phase margin is 45 degrees, the system is considered stable.

Real-world Examples

Aircraft Control Systems

Stability analysis ensures that an aircraft remains stable during flight. Control systems are designed to maintain steady flight despite turbulence and other disturbances.

Example

- An aircraft's autopilot system uses root locus analysis to adjust the control surfaces. The root locus plot shows how the poles move with changes in control surface deflection. By ensuring the poles remain in the left half-plane, the autopilot system maintains stable flight.

Power Systems

Stability in power systems ensures reliable electricity supply. Stability analysis helps prevent voltage collapse and power outages by maintaining the balance between generation and consumption.

Example

- A power grid uses Nyquist criterion to analyze the stability of the system under different loading conditions. The Nyquist plot for the open-loop transfer function of the power grid's control system is used to check for encirclements around the critical point. Ensuring no encirclements indicates that the grid will remain stable under normal and fault conditions.

Robotics

Robots require stable control systems for precise movement and operation. Stability analysis helps design controllers that maintain balance and accuracy.

Example

- A robotic arm uses Bode plots to fine-tune its PID controller. The Bode plot of the open-loop transfer function helps determine the gain and phase margins. By adjusting the PID parameters to achieve sufficient margins, the robotic arm achieves smooth and stable movement while handling delicate tasks.

13.3 Feedback Systems

Feedback systems are fundamental in control engineering and regulate the behavior of dynamic systems. They monitor a system's output and use this information to adjust inputs to achieve the desired performance. By doing so, feedback systems can enhance stability, reduce errors, and improve system performance.

Types of Feedback

Feedback in systems can be broadly classified into two types: positive feedback and negative feedback.

Positive Feedback

Positive feedback occurs when the feedback signal is in phase with the input signal, amplifying the effect of the input. This type of feedback can lead to instability if not properly controlled, as it reinforces deviations from the desired output.

- **Characteristics**
  - Amplifies deviations.
  - Can lead to exponential growth or oscillations.

- Used in applications where amplification or triggering is required.

Example

- **Audio Amplifiers:** In a microphone-speaker system, if the microphone picks up sound from the speaker, it creates a loop that amplifies the sound, leading to a loud, high-pitched noise (feedback loop).

Negative Feedback

Negative feedback occurs when the feedback signal is out of phase with the input signal, counteracting the effect of the input. This type of feedback is used to stabilize systems, reduce errors, and improve accuracy.

- **Characteristics**
  - Reduces deviations.
  - Enhances stability and accuracy.
  - Widely used in control systems to maintain desired output.

Example

- **Thermostat System:** In a heating system, the thermostat measures the room temperature and adjusts the heater output to maintain the set temperature. If the room is too warm, the heater output is reduced, and if it is too cold, the output is increased.

Benefits of Feedback

Feedback systems offer several benefits that make them indispensable in various applications.

- **Stability:** Negative feedback helps stabilize systems by reducing the effect of disturbances and uncertainties.
- **Accuracy:** Feedback systems improve the accuracy of the system output by continuously correcting errors.

- **Robustness:** Feedback can make systems more robust to changes in system parameters and external disturbances.
- **Performance:** Properly designed feedback systems can enhance system performance by optimizing response times and reducing overshoot.

Example

- **Cruise Control in Cars:** Cruise control systems use feedback to maintain a constant vehicle speed. The system adjusts the throttle position based on the difference between the actual speed and the desired speed, ensuring smooth and stable speed control.

Real-world Examples

Operational Amplifiers

Op-amps use negative feedback to control gain and maintain signal integrity. The feedback ensures that the output voltage follows the input voltage with high precision.

Example

- In a voltage follower circuit, the output voltage directly tracks the input voltage, providing a high input impedance and low output impedance.

Automatic Door Systems

Feedback is used to control the opening and closing of automatic doors. Sensors detect the presence of an object or person, and the feedback system adjusts the door's motor to open or close accordingly.

Example

- Supermarket doors that open automatically when a person approaches and close after they pass.

Homeostasis

Biological systems use feedback mechanisms to maintain stable internal conditions. For instance, the human body regulates temperature through sweating and shivering.

Example

- When body temperature rises, the body initiates sweating to cool down, and when it falls, shivering generates heat to warm up.

Phase-Locked Loops (PLLs)

Used in communication systems to synchronize the frequency of a signal with a reference frequency. Feedback ensures that the output signal remains locked to the reference signal, maintaining consistent communication.

Example

- PLLs are used in radio transmitters and receivers to maintain stable frequency modulation and demodulation.

13.4 Controllers and Compensators

Controllers are essential components in control systems that adjust the input to a system to achieve the desired output. They continuously monitor the system's output and compare it with the desired setpoint, making necessary adjustments to minimize the error.

Types of Controllers

Proportional (P) Controllers

Proportional controllers adjust the control input proportional to the error. The control action is based on the current error value.

- **Function:** $u(t) = K_p e(t)$
    - K_p: Proportional gain

- $e(t)$: Error (difference between setpoint and process variable)

- **Advantages:** Simple and easy to implement.
- **Disadvantages:** Can result in steady-state error.
- **Example:** In a temperature control system, a proportional controller adjusts the heater's power based on the difference between the desired temperature and the actual temperature.

Integral (I) Controllers

Integral controllers adjust the control input based on the cumulative sum of past errors. This helps eliminate steady-state error.

- **Function:** $u(t) = K_i \int_0^t e(\tau)d\tau$
  - K_i : Integral gain
- **Advantages:** Eliminates steady-state error.
- **Disadvantages:** Can lead to slow response and potential instability.
- **Example:** In a level control system for a tank, an integral controller adjusts the inlet valve position based on the cumulative difference between the desired and actual water levels.

Derivative (D) Controllers

Derivative controllers adjust the control input based on the rate of change of the error. This anticipates future error and improves system stability.

- **Function:** $u(t) = K_d \frac{de(t)}{dt}$
  - K_d: Derivative gain
- **Advantages:** Improves stability and response speed.

- **Disadvantages:** Sensitive to noise.
- **Example:** In a motor's speed control system, a derivative controller adjusts the motor's power based on the rate of change of the speed error.

PID Controllers PID

controllers combine proportional, integral, and derivative control actions to provide a balanced and robust control response.

- **Function:** $u(t) = K_p e(t) + K_i \int_0^t e(\tau)d\tau + K_d \frac{de(t)}{dt}$
- **Advantages:** Combines the benefits of P, I, and D controllers, providing excellent control performance.
- **Disadvantages:** More complex to tune.
- **Example:** In an industrial automation system, a PID controller regulates the position of a robotic arm by adjusting its motors based on the current position error, cumulative error, and rate of change of the error.

Compensators

Compensators are devices or algorithms added to a control system to improve its performance. They adjust the system's dynamics to achieve desired characteristics such as stability, speed of response, and reduced overshoot.

Lead Compensator

A lead compensator improves system response speed and increases phase margin, making the system more stable.

- **Function:** Adds a zero and a pole to the transfer function, with the zero closer to the origin.
- **Advantages:** Increases system bandwidth and stability.

- **Example:** Used in flight control systems to improve the responsiveness and stability of an aircraft.

Lag Compensator

A lag compensator improves steady-state accuracy and reduces steady-state error.

- **Function:** Adds a pole and a zero to the transfer function, with the pole closer to the origin.
- **Advantages:** Enhances low-frequency gain, improving steady-state performance.
- **Example:** Used in motor control systems to reduce the steady-state error in position control.

Lead-Lag Compensator

A lead-lag compensator combines the benefits of both lead and lag compensators, providing improved transient and steady-state performance.

- **Function:** Adds both lead and lag components to the transfer function.
- **Advantages:** Balances the improvement in stability, response speed, and steady-state accuracy.
- **Example:** Used in automotive cruise control systems to maintain a desired speed with minimal overshoot and steady-state error.

Real-world Examples

Temperature Control in HVAC Systems

- **Controller:** A PID controller adjusts the heating and cooling elements to maintain the desired room temperature.
- **Compensator:** A lead-lag compensator might be used to improve the response time and accuracy of the temperature control.

Robotic Arm Control

- **Controller:** A PID controller regulates the position and movement of the robotic arm.
- **Compensator:** A lead compensator enhances the responsiveness, allowing precise and quick movements.

Automotive Cruise Control

- **Controller:** A PID controller maintains the vehicle's speed by adjusting the throttle.
- **Compensator:** A lag compensator reduces the steady-state error, ensuring accurate speed maintenance.

- VIDEO RESOURCE RECOMMENDATIONS

Enhance your learning with our recommended video resources for a deeper understanding of the concepts. When using these resources, please check the current exam specifications.

An abundance of high-quality video resources is available for preparing for this section. For a comprehensive overview, we recommend the following:

YouTube Playlist: *FE Electrical and Computer: Review (Effective from July 2020) by Amit Roy*
Search for this playlist on YouTube. It covers almost all the exam topics, providing detailed explanations and condensing the material effectively.

Online Review Course: For a more organized approach, consider the comprehensive, free, video-based online review course called *"Electrical FE Review"* by Zach Stone, available at electricalfereview.com

Block Diagrams	
Key Concepts	**Handbook Page References**
Series	
Parallel	
Feedback Loop	

Bode Plots	
Key Concepts	**Handbook Page References**
Magnitude Plot	
Phase Plot	

Closed-Loop and Open-Loop Response	
Key Concepts	Handbook Page References
Routh Test	
Steady State Error:	
- Type 0	
- Type 1	
- Type 2	
Final Value Theorem	
Stability:	
- Gain Margin	
- Phase Margin	
Settling Time	
Overshoot	

Controller Performance	
Key Concepts	Handbook Page References
Proportional Control	
Integral Control	
Derivative Control	

NOTES/EQUATIONS SUMMARY

DIAGNOSTIC TEST CHECKLIST

Checklist Questions	Yes/No
Have you read and understood all the material in this chapter?	
Are you familiar with all the key concepts and formulas highlighted in the tables for this chapter?	
Have you filled in the tables with the correct NCEES® FE Reference Handbook page numbers for each listed concept and formula?	
Did you download and complete all the designated practice problems for this section?	
Did you review and understand the step-by-step solutions to the practice problems?	
Could you solve the practice problems without excessive reference to the solutions and within a reasonable time frame?	
Are you confident in applying the concepts and formulas from this chapter?	
Have you noted any areas of difficulty or topics that require further review before proceeding?	
Do you know how to quickly find information in the NCEES® FE Reference Handbook?	
Have you reviewed any errors or misunderstandings in this section's material to ensure clarity and understanding?	

CHAPTER 14

Communications

Communication systems are integral to the modern world, enabling the transmission of information over various distances and mediums. From traditional analog methods to advanced digital technologies, communication systems have evolved to meet the increasing demand for faster, more reliable, and more efficient information exchange.

In this chapter, we will explore the fundamental concepts and technologies that underpin both analog and digital communication systems. We will discuss how signals are encoded and processed, the types of transmission media used to carry these signals, and the network protocols and architectures that facilitate communication across different platforms and devices.

14.1 Analog and Digital Communication

Communication systems are essential for transmitting information from one place to another. These systems can be broadly classified into analog and digital communication systems. Both types of communication systems have unique characteristics, advantages, and applications. Understanding these differences is crucial for selecting the appropriate system for specific needs.

Analog Communication

Analog communication involves transmitting continuous signals that vary in amplitude or frequency. These signals are analogous to the original information being sent, such as sound or light.

Amplitude Modulation (AM)

Amplitude Modulation is a technique used to encode information by varying the amplitude of a carrier wave. AM radio broadcasts use amplitude modulation to transmit audio signals. The varying amplitude of the carrier wave corresponds to the sound waves of the audio signal, which can be received and demodulated by an AM radio receiver.

- **Principle:** The amplitude of the carrier wave is varied in proportion to the amplitude of the input signal.
- **Advantages**
  - Simple to implement
  - Cost-effective for certain applications
- **Disadvantages**
  - Susceptible to noise and interference
  - Less efficient in terms of bandwidth usage

Frequency Modulation (FM)

Frequency Modulation encodes information by varying the frequency of the carrier wave. FM radio broadcasts use frequency modulation to transmit audio signals. The frequency of the carrier wave changes according to the audio signal, providing clearer sound with less noise compared to AM.

- **Principle:** The frequency of the carrier wave is varied in proportion to the amplitude of the input signal.
- **Advantages**
  - Higher resistance to noise and interference
  - Better sound quality for audio transmissions
- **Disadvantages**
  - More complex and expensive to implement than AM
  - Requires more bandwidth

Digital Communication

Digital communication involves transmitting discrete signals, usually in binary form. These signals represent information using a series of 0s and 1s.

Pulse Code Modulation (PCM)

Pulse Code Modulation (PCM) is a method for digitally representing analog signals. It is widely used in digital telephony and audio recording. An analog voice signal is sampled and converted into a digital signal for transmission over digital communication networks.

- **Principle:** The analog signal is sampled at regular intervals, and each sample is quantized to the nearest value within a range of digital steps.
- **Advantages**
  - High accuracy and quality
  - Less susceptible to noise and interference

- Disadvantages
    - Requires more bandwidth
    - More complex to implement

Digital Modulation Techniques

Digital modulation involves altering a carrier signal based on digital data. Common techniques include Amplitude Shift Keying (ASK), Frequency Shift Keying (FSK), and Phase Shift Keying (PSK).

Amplitude Shift Keying (ASK)

- **Principle:** The amplitude of the carrier wave is changed according to the digital signal.
- **Example:** Used in optical fiber communication where the presence or absence of light represents binary data.

Frequency Shift Keying (FSK)

- **Principle:** The frequency of the carrier wave is varied based on the digital signal.
- **Example:** Used in modem communication for transmitting data over telephone lines.

Phase Shift Keying (PSK)

- **Principle:** The phase of the carrier wave is altered according to the digital signal.
- **Example:** Used in Wi-Fi and Bluetooth communication for robust data transmission.

Comparison and Applications

Comparison

- **Analog Communication**
    - Continuous signals
    - Simple and cost-effective
    - Susceptible to noise and interference
    - Examples: AM and FM radio
- **Digital Communication**
    - Discrete signals (binary)
    - Higher accuracy and quality
    - Less affected by noise and interference
    - Examples: Digital telephony, Wi-Fi, and Bluetooth

Applications

- **Analog Communication**
    - AM and FM radio broadcasting
    - Analog TV transmission
- **Digital Communication**
    - Cellular networks
    - Internet communication
    - Digital TV and radio

- Data communication in computers and smartphones

Example: Digital communication is primarily used in a modern communication system like a smartphone. Voice signals are captured, digitized using PCM, and transmitted over a cellular network using digital modulation techniques like PSK. This ensures clear, reliable communication even in environments with high noise and interference.

14.2 Signal Encoding and Processing

Signal encoding is the process of converting information into a format that can be easily transmitted and decoded at the receiver end. Encoding ensures that the data is efficiently and accurately conveyed over a communication channel. Proper encoding techniques are crucial for minimizing errors and maximizing data integrity during transmission.

Encoding Techniques

Binary Encoding

Binary encoding is the simplest form of encoding where information is represented using two symbols: 0 and 1. Each symbol or bit represents a piece of the data. In computer systems, text characters are often encoded using the ASCII code, where each character is represented by a unique 7-bit or 8-bit binary number.

- **Principle:** Data is converted into a series of binary digits (bits) for transmission.
- **Advantages**
    - Simple and easy to implement
    - Efficient for digital systems

- **Disadvantages**
    - Can be prone to errors in noisy environments

Manchester Encoding

Manchester encoding is a method for encoding binary data to ensure synchronization between the transmitter and receiver. It combines data and clock signals into a single self-synchronizing data stream. Ethernet uses Manchester encoding to ensure data integrity is maintained during transmission over network cables.

- **Principle:** Each bit is represented by a transition. A logical '0' is indicated by a high-to-low transition, and a logical '1' is indicated by a low-to-high transition.
- **Advantages**
    - Provides synchronization without requiring a separate clock signal
    - Reduces the likelihood of errors due to timing issues
- **Disadvantages**
    - Less bandwidth-efficient compared to other encoding schemes

Signal Processing

Filtering

Filtering is the process of removing unwanted components or noise from a signal. Filters can be analog or digital and enhance the quality of the signal. In audio processing, low-pass filters remove high-frequency noise from recordings, enhancing sound quality.

- **Low-pass Filter:** Allows signals with a frequency lower than a certain cutoff frequency to pass through and attenuates frequencies higher than the cutoff.

- **High-pass Filter:** Allows signals with a frequency higher than a certain cutoff frequency to pass through and attenuates frequencies lower than the cutoff.
- **Band-pass Filter:** Allows signals within a certain frequency range to pass through and attenuates frequencies outside this range.

Sampling and Quantization

Sampling and quantization are crucial steps in converting an analog signal into a digital format. In digital audio, an analog sound wave is sampled at a rate of 44.1 kHz and quantized into 16-bit values for CD-quality sound.

- **Sampling:** The process of measuring the amplitude of an analog signal at regular intervals.
    - **Nyquist Theorem:** To accurately reconstruct an analog signal, it must be sampled at least twice the highest frequency present in the signal.
- **Quantization:** The process of mapping the sampled amplitudes to a finite set of values, which are then represented digitally.

Real-world Applications

Digital Telephony

- **Encoding:** Voice signals are encoded using Pulse Code Modulation (PCM) to convert analog voice signals into digital data.
- **Signal Processing:** Filters are used to remove background noise, and the signal is sampled and quantized for transmission over digital networks.

Wireless Communication

- **Encoding:** Data is encoded using binary or more complex encoding schemes like Quadrature Amplitude Modulation (QAM) for efficient transmission.

- **Signal Processing:** Filters and equalizers are used to mitigate the effects of interference and multipath fading, ensuring clear signal reception.

Medical Imaging

- **Encoding:** Imaging data from devices like MRI or CT scans is encoded digitally for storage and analysis.
- **Signal Processing:** Filters and other processing techniques enhance image quality by removing noise and improving contrast.

Broadcasting

- **Encoding:** Television and radio signals are encoded using standards like MPEG for video and AAC for audio.
- **Signal Processing:** Compression and filtering techniques are used to optimize bandwidth usage and improve signal quality.

14.3 Transmission Media

Transmission media are the physical pathways through which data is transmitted from one location to another. They can be broadly classified into wired and wireless media.

Wired Media

Twisted Pair

Consists of pairs of insulated copper wires twisted together to reduce electromagnetic interference. UTP cables are used in local area networks (LANs) to connect computers and network devices.

- **Types**
  - **Unshielded Twisted Pair (UTP):** Commonly used in telephone and Ethernet networks.

- **Shielded Twisted Pair (STP):** Provides additional shielding to reduce interference further.

Coaxial Cable

Consists of a central conductor, an insulating layer, a metallic shield, and an outer insulating layer. Coaxial cables are used to connect televisions to cable service providers, providing high-quality video signals.

- **Applications:** Used for cable television, internet connections, and some Ethernet networks.

Optical Fiber

Uses light signals to transmit data through strands of glass or plastic fibers. Optical fibers are used in high-speed internet connections, providing fast and reliable data transmission over long distances.

- Types
    - **Single-mode Fiber:** Supports long-distance communication with a single light mode.
    - **Multi-mode Fiber:** Supports shorter distances with multiple light modes.

Wireless Media

Radio Waves

Uses electromagnetic waves in the radio frequency range to transmit data. Wi-Fi networks use radio waves to provide wireless internet access within a specific area, such as a home or office.

- **Applications:** Used for AM/FM radio, television broadcasts, and wireless networking (Wi-Fi).

Microwaves

Uses high-frequency radio waves to transmit data over long distances with line-of-sight propagation. Microwave communication links are used to transmit data between mobile phone towers.

- **Applications:** Used for satellite communication, cellular networks, and point-to-point communication links.

Infrared

Uses infrared light to transmit data over short distances. Infrared is used in TV remote controls to send signals to the television set.

- **Applications:** Used for remote controls, short-range communication between devices, and some wireless local area networks.

Characteristics and Applications

- **Bandwidth:** The amount of data that can be transmitted over a medium in a given time. Optical fiber has the highest bandwidth, followed by coaxial cable, twisted pair, and wireless media.

- **Distance:** Wired media like optical fiber and coaxial cable support long-distance communication, while wireless media like radio waves and microwaves are suitable for both short and long distances, depending on the application.

- **Interference:** Wired media are generally less susceptible to interference compared to wireless media. Optical fiber is immune to electromagnetic interference.

- **Installation and Cost:** Wired media can be more expensive and difficult to install, especially over long distances. Wireless media offer easier installation and flexibility.

Advantages and Disadvantages

Twisted Pair

- **Advantages:** Low cost, easy to install, and widely available.
- **Disadvantages:** Limited bandwidth and distance, susceptible to electromagnetic interference.

Coaxial Cable

- **Advantages:** Higher bandwidth than twisted pair, better shielding against interference.
- **Disadvantages:** Bulkier and more expensive than twisted pair, more difficult to install.

Optical Fiber

- **Advantages:** Very high bandwidth, long-distance communication, immune to electromagnetic interference.
- **Disadvantages:** Expensive and fragile, requires specialized equipment for installation and maintenance.

Radio Waves

- **Advantages:** Wireless and flexible, supports mobility, wide coverage area.
- **Disadvantages:** Susceptible to interference and signal degradation, limited bandwidth compared to wired media.

Microwaves

- **Advantages:** High bandwidth, suitable for long-distance communication, point-to-point links.
- **Disadvantages:** Requires line-of-sight, can be affected by weather conditions, more expensive.

Infrared

- **Advantages:** Secure, not susceptible to electromagnetic interference, low cost.
- **Disadvantages:** Limited range and line-of-sight required, can be blocked by obstacles.

14.4 Network Protocols and Architectures

Network protocols are rules and conventions that govern how data is transmitted and received across networks. These protocols ensure that devices from different manufacturers can communicate effectively and reliably.

- **Purpose:** Facilitate communication between devices over a network.
- **Examples:** Protocols define how data is formatted, transmitted, and processed, ensuring interoperability.

OSI and TCP/IP Models

The OSI (Open Systems Interconnection) model and TCP/IP (Transmission Control Protocol/Internet Protocol) model are frameworks used to understand and design network protocols.

OSI Model

- **Layers:** The OSI model has seven layers - Physical, Data Link, Network, Transport, Session, Presentation, and Application.
- **Purpose:** Each layer performs specific functions to support communication across networks, from physical transmission to application-level interactions.
- **Example:** The Data Link layer ensures error-free transmission between adjacent network nodes, while the Transport layer provides end-to-end communication services for applications.

TCP/IP Model

- **Layers:** The TCP/IP model has four layers - Network Interface, Internet, Transport, and Application.

- **Purpose:** Simplified model used primarily for the internet, focusing on how data is transmitted from one network to another.

- **Example:** The Internet layer (IP) is responsible for addressing and routing packets of data, while the Transport layer (TCP) ensures reliable data transmission.

Common Network Protocols

HTTP (HyperText Transfer Protocol)

- **Purpose:** Used for transferring web pages on the internet.

- **Function:** Defines how messages are formatted and transmitted, and how web servers and browsers should respond to commands.

- **Example:** When you enter a URL in a web browser, HTTP is used to request the web page from a server and display it on your screen.

FTP (File Transfer Protocol)

- **Purpose:** Used for transferring files between computers on a network.

- **Function:** Allows users to upload and download files from remote servers.

- **Example:** A company might use FTP to distribute software updates to customers.

TCP (Transmission Control Protocol)

- **Purpose:** Ensures reliable transmission of data over a network.

- **Function:** Establishes a connection between two devices and ensures data is delivered in the correct order without errors.

- **Example:** Email applications use TCP to send messages, ensuring that each email reaches its destination intact.

IP (Internet Protocol)

- **Purpose:** Handles addressing and routing of packets across networks.
- **Function:** Defines how data packets are formatted and transmitted from one device to another.
- **Example:** IP addresses are used to identify devices on a network, ensuring data reaches the correct destination.

Network Architectures

Client-Server

In a client-server architecture, a server provides resources or services, and clients access these resources or services. A web server hosting a website and multiple users accessing the site from their browsers.

- **Characteristics**
    - Centralized control
    - Easier management of resources
    - Can handle large networks

Peer-to-Peer

In a peer-to-peer (P2P) architecture, each device on the network can act as both a client and a server. File-sharing applications like BitTorrent, where users download and upload files directly from/to each other.

- **Characteristics**
    - Decentralized control

- Equal sharing of resources
- More difficult to manage in large networks

Applications in Modern Communications Systems

Network protocols and architectures are fundamental to the operation of modern communication systems. They enable the efficient and reliable transmission of data across different platforms and devices.

Example

- **Internet:** Uses the TCP/IP model with protocols like HTTP, FTP, TCP, and IP to facilitate web browsing, file transfers, and data communication.
- **Enterprise Networks:** Often use client-server architecture for resource management, with protocols like TCP and IP ensuring reliable data transmission.
- **Smart Home Systems:** Utilize a mix of client-server and peer-to-peer architectures, with protocols like HTTP and IP enabling device communication and control.

ADDITIONAL ONLINE RESOURCES

- **1500+ PRACTICE PROBLEMS [4 PRACTICE EXAMS INCLUDED]**

Enhance your exam preparation with over 1000+ practice problems, complete with step-by-step solutions. Two weeks after your initial download, you will receive a weekly full-length practice exam with step-by-step solutions for four weeks, totaling over 1500 practice problems.

SCAN HERE

Having trouble accessing the resources? Email us at **info@learnovapublishing.com** — we reply within 24 hours, 7 days a week

- **VIDEO RESOURCE RECOMMENDATIONS**

Enhance your learning with our recommended video resources for a deeper understanding of the concepts. When using these resources, please check the current exam specifications.

An abundance of high-quality video resources is available for preparing for this section. For a comprehensive overview, we recommend the following:

YouTube Playlist: *FE Electrical and Computer: Review (Effective from July 2020) by Amit Roy*
Search for this playlist on YouTube. It covers almost all the exam topics, providing detailed explanations and condensing the material effectively.

Online Review Course: For a more organized approach, consider the comprehensive, free, video-based online review course called *"Electrical FE Review"* by Zach Stone, available at electricalfereview.com

Basic Modulation and Demodulation	
Key Concepts	**Handbook Page References**
Amplitude Modulation (AM):	
- Amplitude Modulation Efficiency	
- Power Transmission for a Message	
- Modulation Index	
Double-Sideband Modulation (DSB)	
Single-Sideband Modulation (SSB)	
Angle Modulation:	
- Frequency Modulation (FM)	
- 98% Bandwidth Power of FM Signal	
- Phase Modulation (PM)	

Fourier Transform and Fourier Series	
Key Concepts	**Handbook Page References**
Common Fourier Transform Pairs:	
- Delta	
- Unit Step	
- Square Pulse (Box)	
- Triangle	
- Exponential	
- Cosine	
- Sine	
Fourier Transform Properties:	

- Linearity	
- Frequency Shifting	
- Time Scaling	
- Time Differentiation & Time Integration	
- Convolution	

Multiplexing	
Key Concepts	**Handbook Page References**
Time Division Multiplexing (TDM)	
Frequency Division Multiplexing (FDM)	
Code Division Multiplexing (CDM)	

Digital Communications	
Key Concepts	**Handbook Page References**
Pulse Width Modulation (PWM)	
Pulse Amplitude Modulation (PAM)	
Pulse Code Modulation (PCM)	
Parity Bit Coding	
Cyclical Redundancy Code (CRC)	

NOTES/EQUATIONS SUMMARY

DIAGNOSTIC TEST CHECKLIST

Checklist Questions	Yes/No
Have you read and understood all the material in this chapter?	
Are you familiar with all the key concepts and formulas highlighted in the tables for this chapter?	
Have you filled in the tables with the correct NCEES® FE Reference Handbook page numbers for each listed concept and formula?	
Did you download and complete all the designated practice problems for this section?	
Did you review and understand the step-by-step solutions to the practice problems?	
Could you solve the practice problems without excessive reference to the solutions and within a reasonable time frame?	
Are you confident in applying the concepts and formulas from this chapter?	
Have you noted any areas of difficulty or topics that require further review before proceeding?	
Do you know how to quickly find information in the NCEES® FE Reference Handbook?	
Have you reviewed any errors or misunderstandings in this section's material to ensure clarity and understanding?	

CHAPTER 15

Computer Networks

Computer networks are the backbone of modern communication, enabling devices to connect and share information seamlessly. From personal devices like smartphones and laptops to large-scale enterprise systems, networks facilitate data exchange, resource sharing, and connectivity across the globe. Understanding the fundamental concepts of computer networks is essential for designing, implementing, and managing these systems efficiently.

In this chapter, we will explore the core principles and techniques that underpin computer networks. We will delve into the various types of network topologies and devices, examine how data communication is achieved through different protocols and transmission modes, and discuss the crucial aspects of network security. Additionally, we will cover the impact of internet technologies, including the World Wide Web (WWW), the Internet of Things (IoT), and cloud computing, highlighting their applications in real-world scenarios.

15.1 Network Principles

Network principles are the foundational concepts that govern the design, implementation, and operation of computer networks. These principles ensure that networks function efficiently, allowing devices to communicate and share resources seamlessly. Understanding network principles is essential for anyone involved in networking, as they form the basis for building robust and scalable communication systems.

Network Topologies

Network topology refers to the arrangement of different elements (links, nodes) in a computer network. It defines how devices are interconnected and how data flows between them. There are several types of network topologies, each with its advantages and disadvantages.

Star Topology

In a star topology, all devices are connected to a central hub or switch. The hub acts as a repeater for data flow. A typical home network where all devices (computers, printers) connect to a central router.

- **Advantages**
  - Easy to install and manage.
  - Failure of one device does not affect the others.
- **Disadvantages**
  - If the central hub fails, the entire network is down.

Bus Topology

In a bus topology, all devices share a single communication line or bus. Each device is connected to this central cable. Early Ethernet networks used a bus topology to connect multiple devices on a single cable.

- **Advantages**
    - Simple and cost-effective for small networks.
    - Easy to extend by adding more devices to the bus.
- **Disadvantages**
    - If the main cable fails, the entire network is affected.
    - Performance degrades as more devices are added.

Ring Topology

In a ring topology, each device is connected to two other devices, forming a circular data path. Data travels in one direction. Token Ring networks used in some business environments to manage data traffic efficiently.

- **Advantages**
    - Data packets travel quickly due to the consistent data path.
    - Easy to identify and isolate faults.
- **Disadvantages**
    - Failure of one device can disrupt the entire network.
    - Adding or removing devices can be complex.

Mesh Topology

In a mesh topology, each device is connected to every other device. This creates multiple paths for data to travel. Military and mission-critical networks where reliability and redundancy are paramount.

- **Advantages**
    - Highly reliable due to redundant paths.
    - Easy to identify and isolate faults without affecting the entire network.

- **Disadvantages**
    - Expensive and complex to install and maintain.
    - Requires a lot of cabling and configuration.

Network Devices

Network devices are the hardware components that facilitate communication between networked devices. They play a crucial role in directing data traffic and ensuring efficient data transmission.

Routers

Routers connect different networks and direct data packets between them. They use IP addresses to determine the best path for data. A router connects a home network to the internet, directing traffic between the local devices and external networks.

- **Advantages**
    - Can connect different network types.
    - Provides security features like firewalls.
- **Disadvantages**
    - More complex and expensive than switches or hubs.

Switches

Switches connect devices within the same network, using MAC addresses to forward data to the correct destination. An Ethernet switch connects computers, printers, and other devices in an office, allowing them to communicate efficiently.

- **Advantages**
    - Efficient data handling within a local network.
    - Reduces collisions and improves performance.

- **Disadvantages**
    - Does not route data between different networks.

Hubs

Hubs connect multiple Ethernet devices, making them act as a single network segment. They broadcast incoming data to all ports. A hub connecting several computers in a small home network, allowing basic communication.

- **Advantages**
    - Simple and inexpensive.
    - Easy to install and use.
- **Disadvantages**
    - Inefficient as it broadcasts data to all devices, leading to potential collisions and reduced performance.

Real-world Examples

Enterprise Networks

Large organizations often use a combination of topologies and devices to create a robust network infrastructure. For example, a star topology might be used for office workstations connected to a central switch, while a mesh topology ensures redundancy and reliability for critical servers.

Educational Institutions

Schools and universities typically use a bus or star topology for classroom and administrative networks. Switches and routers manage data traffic, ensuring students and staff have reliable internet access and internal communication.

Home Networks

A typical home network setup involves a router connecting to an internet service provider, with various devices (smartphones, laptops, smart TVs) connected via a star topology. A switch may be used to expand the number of wired connections available.

15.2 Data Communication

Data communication refers to the exchange of data between devices through a transmission medium such as wires, cables, or wireless signals. It encompasses the processes, protocols, and technologies that ensure data is accurately transmitted, received, and understood by the communicating devices. Effective data communication is crucial for the functioning of computer networks, enabling the sharing of resources and information.

Communication Protocols

Communication protocols are sets of rules that determine how data is transmitted and received over a network. They ensure that devices can communicate with each other effectively, despite differences in their underlying hardware and software.

TCP/IP (Transmission Control Protocol/Internet Protocol)

TCP/IP is a suite of communication protocols used to interconnect network devices on the internet. TCP ensures reliable, ordered, and error-checked delivery of data, while IP handles addressing and routing. When you send an email, TCP divides the message into packets and ensures they are delivered and reassembled correctly, while IP ensures the packets reach the correct recipient.

Function

- **TCP**: Establishes a connection between sender and receiver, breaks data into packets, ensures all packets are received and reassembled correctly.
- **IP**: Addresses and routes each packet to its destination.

UDP (User Datagram Protocol)

UDP is a simpler, connectionless protocol that does not guarantee delivery, order, or error checking. It is used for applications where speed is more critical than reliability. Online gaming and live video streaming use UDP to achieve low latency, even if some data packets are lost or received out of order.

Function

- Sends data packets without establishing a connection.
- Does not ensure packets are received or reassembled in order.

Data Transmission Modes

Data transmission modes refer to the methods used to send data between devices. These modes determine how and when data is transmitted.

Synchronous Transmission

Data is transmitted at regular intervals, synchronized by a clock signal. Both sender and receiver must be synchronized. Synchronous transmission is used in high-speed data networks like Ethernet, where devices are synchronized to ensure efficient data transfer.

- **Advantages**
  - High-speed data transfer
  - Suitable for large data volumes
- **Disadvantages**
  - Requires precise synchronization
  - More complex and expensive

Asynchronous Transmission

Data is transmitted one byte at a time, each byte being framed by start and stop bits. There is no need for synchronization between sender and receiver. Asynchronous transmission is commonly used in serial communication, such as between a computer and a modem, where data is sent one byte at a time.

- **Advantages**
  - Simple and cost-effective
  - No need for precise timing
- **Disadvantages**
  - Slower data transfer
  - Overhead due to start and stop bits

Real-world Examples

Web Browsing

- **Protocol:** Uses TCP/IP to ensure reliable data transfer. When you access a website, your browser sends HTTP requests over TCP/IP, ensuring the complete and correct delivery of web pages.
- **Transmission Mode:** Primarily uses synchronous transmission for fast and efficient data transfer between servers and clients.

Video Streaming

- **Protocol:** Often uses UDP to reduce latency. Streaming services like Netflix and YouTube use UDP to deliver video content quickly, even if it means some data packets might be lost.
- **Transmission Mode:** Uses a combination of synchronous and asynchronous modes to balance speed and reliability.

Email

- **Protocol:** Uses TCP/IP for reliable data transfer. Email clients use SMTP (Simple Mail Transfer Protocol) for sending emails and IMAP (Internet Message Access Protocol) or POP3 (Post Office Protocol) for retrieving emails.
- **Transmission Mode:** Synchronous transmission ensures that email data is sent and received efficiently and in the correct order.

15.3 Network Security

Network security involves protecting network infrastructure and data from unauthorized access, misuse, and cyber threats. It is crucial for maintaining the integrity, confidentiality, and availability of information in computer networks. With the increasing reliance on digital communication, ensuring robust network security measures is essential to safeguard sensitive data and maintain user trust.

Types of Network Security Threats

Malware

Malware is malicious software designed to damage, disrupt, or gain unauthorized access to computer systems. It includes viruses, worms, trojans, ransomware, and spyware.

- **Impact:** Can lead to data loss, system damage, and unauthorized data access.
- **Example:** A computer virus that infects a network and corrupts files, causing operational disruptions and data breaches.

Phishing

Phishing involves tricking individuals into providing sensitive information, such as usernames, passwords, and credit card details, by posing as a trustworthy entity.

- **Impact:** Can lead to identity theft, financial loss, and unauthorized access to sensitive information.
- **Example:** An email that appears to be from a bank asking users to click on a link and enter their login details, leading to stolen credentials.

Denial-of-Service (DoS)

A DoS attack aims to make a network service unavailable by overwhelming it with a flood of illegitimate requests, causing system overload.

- **Impact:** Can disrupt services, cause downtime, and result in significant financial losses.
- **Example:** A website becomes inaccessible due to a DoS attack that floods the server with excessive traffic, preventing legitimate users from accessing the site.

Security Measures

Firewalls

Firewalls are network security devices or software that monitor and control incoming and outgoing network traffic based on predefined security rules.

- **Function:** Act as a barrier between trusted internal networks and untrusted external networks, blocking unauthorized access.
- **Example:** A firewall prevents external attackers from accessing a company's internal network by blocking suspicious traffic based on security policies.

Encryption

Encryption converts data into a coded format that can only be read by someone with the correct decryption key. It ensures data confidentiality and integrity.

- **Function:** Protects data during transmission and storage, preventing unauthorized access.
- **Example:** Secure Socket Layer (SSL) encryption is used in online banking to encrypt sensitive financial data, ensuring that only the intended recipient can read the information.

Intrusion Detection Systems (IDS)

IDS are systems that monitor network traffic for suspicious activity and potential security breaches. They can be network-based (NIDS) or host-based (HIDS).

Function: Detects and alerts administrators about possible security incidents, allowing for timely response.

- **Example:** An IDS identifies unusual traffic patterns that indicate a potential attack, such as multiple failed login attempts, and alerts the network administrator to take action.

Real-world Examples

Corporate Networks

- **Threat:** A large corporation faces a malware attack that infects its network and encrypts critical data, demanding a ransom for decryption.
- **Security Measure:** The company uses advanced firewalls, encryption, and IDS to detect and block the attack, preventing data loss and minimizing downtime.

E-commerce Websites

- **Threat:** An e-commerce website experiences a phishing attack where customers receive fake emails asking for their login credentials.
- **Security Measure:** The website implements multi-factor authentication (MFA) and educates customers about phishing, reducing the risk of credential theft.

Government Agencies

- **Threat:** A government agency faces a DoS attack aimed at disrupting its online services.
- **Security Measure:** The agency deploys firewalls and IDS to detect and mitigate the attack, ensuring continuous service availability and protecting critical infrastructure.

15.4 Internet Technologies

Internet technologies encompass the various protocols, tools, and frameworks that enable communication, data exchange, and connectivity over the internet. These technologies have transformed how we live, work, and interact, making information accessible and fostering innovation across multiple fields. Understanding these technologies is crucial for leveraging the full potential of the internet.

World Wide Web (WWW)

The World Wide Web is a system of interlinked hypertext documents and multimedia content accessed via the internet. It allows users to navigate and access information using web browsers.

Components

- **Web Pages:** Documents written in HTML that can include text, images, videos, and links.
- **Web Browsers:** Software applications that retrieve, display, and navigate web pages (e.g., Chrome, Firefox).
- **Web Servers:** Computers that store and serve web content to users.
- **Example:** When you search for information on Google, the web browser sends a request to Google's servers, which then return relevant web pages for the browser to display.

Internet of Things (IoT)

The Internet of Things refers to the network of physical objects embedded with sensors, software, and other technologies to connect and exchange data with other devices and systems over the internet.

Components

- **Sensors and Actuators:** Collect data from the environment and perform actions based on that data.
- **Connectivity:** Technologies like Wi-Fi, Bluetooth, and cellular networks that enable devices to communicate.
- **Data Processing:** Analyzing and processing data collected from devices to make informed decisions.
- **Example:** A smart home system where devices like thermostats, lights, and security cameras are connected to the internet, allowing homeowners to control and monitor their home remotely via a smartphone app.

Cloud Computing

Cloud computing provides on-demand delivery of computing resources over the internet. It allows users to access and store data, run applications, and utilize IT infrastructure without managing physical servers.

Service Models

- **Infrastructure as a Service (IaaS):** Provides virtualized computing resources over the internet (e.g., Amazon Web Services).
- **Platform as a Service (PaaS):** Provides a platform allowing customers to develop, run, and manage applications (e.g., Google App Engine).
- **Software as a Service (SaaS):** Delivers software applications over the internet (e.g., Microsoft Office 365).

- **Example:** Using Google Drive to store and share documents, which are accessible from any device with an internet connection.

Real-world Examples

Healthcare

- **WWW:** Medical professionals access online databases and journals to stay updated with the latest research and treatments.
- **IoT:** Wearable health monitors collect patient data and transmit it to healthcare providers for real-time monitoring and analysis.
- **Cloud Computing:** Hospitals use cloud-based electronic health records (EHR) systems to store patient information securely and access it from anywhere.

Education

- **WWW:** Students and educators use educational websites and online resources for learning and teaching.
- **IoT:** Smart classrooms equipped with IoT devices like interactive whiteboards and connected projectors enhance the learning experience.
- **Cloud Computing:** Educational institutions use cloud services for hosting virtual classrooms, managing student data, and running applications like learning management systems (LMS).

Retail

- **WWW:** E-commerce platforms allow consumers to shop online, accessing a wide range of products and services.
- **IoT:** Retailers use IoT devices for inventory management, tracking stock levels, and optimizing supply chains.

- **Cloud Computing:** Retail companies use cloud-based CRM systems to manage customer relationships and analyze purchasing trends.

ADDITIONAL ONLINE RESOURCES

- **1500+ PRACTICE PROBLEMS [4 PRACTICE EXAMS INCLUDED]**

Enhance your exam preparation with over 1000+ practice problems, complete with step-by-step solutions. Two weeks after your initial download, you will receive a weekly full-length practice exam with step-by-step solutions for four weeks, totaling over 1500 practice problems.

SCAN HERE

Having trouble accessing the resour[illegible] us at **info@learnovapublishing.com** — we reply within 24 hours, 7 days a week

- **VIDEO RESOURCE RECOMMENDATIONS**

Enhance your learning with our recommended video resources for a deeper understanding of the concepts. When using these resources, please check the current exam specifications.

An abundance of high-quality video resources is available for preparing for this section. For a comprehensive overview, we recommend the following:

YouTube Playlist: *Computer Networks by FE Exam*

Search for this playlist on YouTube. It provides detailed explanations condensing the material effectively.

Online Review Course: For a more organized approach, consider the comprehensive, free, video-based online review course called *"Electrical FE Review"* by Zach Stone, available at electricalfereview.com

Routing and Switching	
Key Concepts	**Handbook Page References**
TCP/IP Model	
OSI Model	
Data Forwarding	
Routing Tables	

Network Topologies	
Key Concepts	**Handbook Page References**
Point to Point	
Token Ring	
Bus	
Star	
Mesh	
Tree	

Network Types	
Key Concepts	**Handbook Page References**
Local Area Networks (LAN):	
- Dynamic Host Configuration Protocol (DHCP)	
- Static IP	
- Stateless Address Auto Configuration (SLAAC)	
Wide Area Network (WAN)	
Internet	

Network Security	
Key Concepts	Handbook Page References
Firewalls	
Port Scanning	
Web Vulnerability Testing	
Penetration Testing	
Encryption Basics	
Diffie-Hellman Key-Exchange Protocol	
RSA Public-Key Cryptosystem	

Network Performance and Error Handling	
Key Concepts	Handbook Page References
Error Coding	
Delays in Computer Networks:	
- Transmission Delay	
- Propagation Delay	
- Queuing Delay	
- Processing Delay	
Automatic Request for Retransmission (ARQ):	
- Stop & Wait	
- Go Back	
- Selective Repeat	
McCabe's Cyclomatic Complexity	

NOTES/EQUATIONS SUMMARY

DIAGNOSTIC TEST CHECKLIST

Checklist Questions	Yes/No
Have you read and understood all the material in this chapter?	
Are you familiar with all the key concepts and formulas highlighted in the tables for this chapter?	
Have you filled in the tables with the correct NCEES® FE Reference Handbook page numbers for each listed concept and formula?	
Did you download and complete all the designated practice problems for this section?	
Did you review and understand the step-by-step solutions to the practice problems?	
Could you solve the practice problems without excessive reference to the solutions and within a reasonable time frame?	
Are you confident in applying the concepts and formulas from this chapter?	
Have you noted any areas of difficulty or topics that require further review before proceeding?	
Do you know how to quickly find information in the NCEES® FE Reference Handbook?	
Have you reviewed any errors or misunderstandings in this section's material to ensure clarity and understanding?	

CHAPTER 16

Digital Systems

Digital systems form the backbone of modern electronic devices and technologies. From the simplest calculators to the most advanced computers and communication systems, digital systems are essential for processing and managing information. Understanding digital systems involves learning about number systems, microprocessors, microcontrollers, memory technologies, logic design, and integrated circuits.

In this chapter, we will explore the fundamental concepts of digital systems, providing both theoretical insights and practical knowledge essential for mastering the digital sections of the FE exam. We will break down complex topics into manageable sections, ensuring clear and straightforward explanations.

16.1 Number Systems and Conversions

Number systems are methods of representing numbers in different forms. In digital systems, the most common number systems are binary, decimal, octal, and hexadecimal. Each system has its own base value, which determines how numbers are represented and manipulated.

- **Decimal (Base 10):** The standard number system used in everyday life, consisting of digits 0-9.
- **Binary (Base 2):** Used in digital electronics and computer systems, consisting of digits 0 and 1.
- **Octal (Base 8):** Uses digits 0-7 and is sometimes used in computing as a shorthand for binary.
- **Hexadecimal (Base 16):** Uses digits 0-9 and letters A-F, commonly used in programming and digital systems for compact representation of binary values.

Binary, Decimal, Octal, and Hexadecimal Systems

Binary System (Base 2)

- **Digits:** 0, 1
- **Usage:** Digital electronics, computer systems
- **Example:** The binary number 1010 represents 10 in decimal.

Decimal System (Base 10)

- **Digits:** 0-9
- **Usage:** Everyday arithmetic and counting
- **Example:** The decimal number 42 remains 42 in decimal notation.

Octal System (Base 8)

- **Digits:** 0-7
- **Usage:** Shorthand notation for binary numbers
- **Example:** The octal number 52 represents 42 in decimal.

Hexadecimal System (Base 16)

- **Digits:** 0-9, A-F
- **Usage:** Compact representation of binary values in computing
- **Example:** The hexadecimal number 2A represents 42 in decimal.

Conversion Techniques

Binary to Decimal

To convert a binary number to decimal, sum the products of each binary digit and its corresponding power of 2.

Steps

1. Write down the binary number.
2. Multiply each digit by 2 raised to the power of its position, starting from 0.
3. Sum the results.

Example

- Binary: 1010
- Calculation: $1 \times 2^3 + 0 \times 2^2 + 1 \times 2^1 + 0 \times 2^0 = 8 + 0 + 2 + 0 = 10$
- Decimal: 10

Decimal to Binary

To convert a decimal number to binary, repeatedly divide the number by 2 and record the remainders.

Steps

1. Divide the decimal number by 2.
2. Write down the remainder.

3. Repeat until the quotient is 0.
4. Read the remainders in reverse order.

Example

- Decimal: 10
- Calculation: $10 \div 2 = 5$ (remainder 0), $5 \div 2 = 2$ (remainder 1), $2 \div 2 = 1$ (remainder 0), $1 \div 2 = 0$ (remainder 1)
- Binary: 1010

Binary to Octal and Hexadecimal

To convert binary to octal, group the binary digits into sets of three, starting from the right, and convert each group to its octal equivalent. For hexadecimal, group the binary digits into sets of four.

Steps for Binary to Octal

1. Group binary digits into sets of three from right to left.
2. Convert each group to its octal equivalent.

Example

- Binary: 101011
- Grouped: 101 011
- Octal: 5 3
- Result: 53

Steps for Binary to Hexadecimal

1. Group binary digits into sets of four from right to left.
2. Convert each group to its hexadecimal equivalent.

Example

- Binary: 101011
- Grouped: 0010 1011
- Hexadecimal: 2B
- Result: 2B

Octal and Hexadecimal to Binary

To convert octal to binary, convert each octal digit to its three-bit binary equivalent. For hexadecimal, convert each hexadecimal digit to its four-bit binary equivalent.

Steps for Octal to Binary

1. Convert each octal digit to its three-bit binary equivalent.

Example

- Octal: 53
- Binary: 101 011
- Result: 101011

Steps for Hexadecimal to Binary

1. Convert each hexadecimal digit to its four-bit binary equivalent.

Example

- Hexadecimal: 2B
- Binary: 0010 1011
- Result: 00101011

Real-world Examples

Digital Systems

- **Usage:** Binary numbers are used in all digital systems, from simple calculators to complex computers, to represent and manipulate data.
- **Example:** A computer's central processing unit (CPU) uses binary to perform all operations. Instructions and data are encoded in binary and processed using binary arithmetic.

Programming

- **Usage:** Hexadecimal numbers are often used in programming to simplify the representation of binary-coded values.
- **Example:** In web design, colors are represented using hexadecimal values, such as #FFFFFF for white.

Networking

- **Usage:** IP addresses in computer networks are often represented in binary for routing and addressing.
- **Example:** The IP address 192.168.1.1 can be represented in binary as 11000000.10101000.00000001.00000001.

16.2 Microprocessors and Microcontrollers

Microprocessors and microcontrollers are essential components in modern electronic systems. A microprocessor is a central processing unit (CPU) on a single chip that performs arithmetic and logic operations. It is the brain of computers, handling complex tasks and running applications. Microcontrollers, on the other hand, are compact integrated circuits designed to govern specific operations in embedded systems. They contain a CPU, memory, and peripherals all in one package, making them ideal for controlling small, dedicated devices.

Architecture of Microprocessors

Microprocessors consist of several key components that work together to process instructions and manage data.

Arithmetic Logic Unit (ALU)

Performs arithmetic and logical operations. It can add, subtract, multiply, divide, and perform logical operations like AND, OR, and NOT.

Central to the processor's ability to perform calculations and make decisions.

Control Unit

Directs the operation of the processor. It fetches instructions from memory, decodes them, and executes them by coordinating the activities of the ALU, registers, and other components.

Ensures instructions are processed correctly and efficiently.

Registers

Small, fast storage locations within the CPU. They temporarily hold data and instructions that the processor is currently working on.

- **Types:**
    - **General-purpose registers:** Hold intermediate data during computation.
    - **Special-purpose registers:** Include the program counter (PC) and status register, which keep track of execution flow and the state of the processor.

Example

In an Intel x86 microprocessor, the ALU performs calculations, the control unit manages the execution of instructions, and registers store temporary data and control information.

Architecture of Microcontrollers

Microcontrollers integrate several components on a single chip to perform dedicated control tasks efficiently.

Integrated Peripherals

Microcontrollers come with built-in peripherals such as analog-to-digital converters (ADCs), digital-to-analog converters (DACs), and pulse-width modulation (PWM) modules.

These peripherals allow the microcontroller to interface with sensors, actuators, and other external devices.

Timers

Timers are built-in features that keep track of time intervals. Used for tasks like generating delays, measuring time intervals, and creating precise output signals.

Communication Ports

Ports like UART, SPI, and I2C allow the microcontroller to communicate with other devices.

Facilitate data exchange between the microcontroller and external components, such as sensors, displays, and other microcontrollers.

Example

The Arduino Uno microcontroller includes an ATmega328P chip with built-in peripherals for analog input, digital input/output, timers, and serial communication ports (UART, I2C, SPI).

Comparison and Applications

Microprocessors

- **Complexity:** Generally more complex and powerful than microcontrollers.

- **Usage:** Found in general-purpose computers, laptops, and high-performance applications.
- **Applications:** Running operating systems, executing software applications, data processing.

Microcontrollers

- **Complexity:** Simpler and more specialized than microprocessors.
- **Usage:** Used in embedded systems for specific control tasks.
- **Applications:** Home appliances, automotive control systems, medical devices, robotics.

Comparison

- **Microprocessors:** Require external components like RAM and ROM for operation. Suitable for tasks requiring high processing power and flexibility.
- **Microcontrollers:** Include CPU, memory, and peripherals on a single chip. Ideal for cost-effective, energy-efficient control applications.

Real-world Examples

Microprocessors

- **Personal Computers:** Intel Core i7 processors in desktop computers handle complex computations, run multiple applications simultaneously, and support high-end gaming and multimedia tasks.
- **Servers:** AMD EPYC processors in data centers manage large-scale data processing, virtualization, and cloud computing services.

Microcontrollers

- **Home Automation:** An ESP8266 microcontroller in a smart thermostat controls the temperature based on user preferences and environmental data.

- **Automotive:** The ATmega16 microcontroller in car engine control units (ECUs) manages fuel injection, ignition timing, and other engine parameters for optimal performance.

- **Wearable Devices:** A microcontroller in a fitness tracker collects sensor data (e.g., heart rate, steps) and processes it to provide real-time feedback to the user.

16.3 Memory Technology and Systems

Memory technology is a crucial aspect of digital systems, responsible for storing and retrieving data and instructions. It plays a vital role in determining the performance and efficiency of computing systems. Memory can be categorized based on its volatility, speed, and usage in the system, ranging from fast, temporary storage to slower, permanent storage.

Types of Memory

Memory can be broadly classified into RAM (Random Access Memory) and ROM (Read-Only Memory), each serving distinct purposes in digital systems.

RAM (Random Access Memory)

- **Volatile Memory:** Loses its contents when power is turned off.

- **Usage:** Used for temporary storage of data that the CPU needs during operation.

SRAM (Static RAM)

- **Description:** Uses bistable latching circuitry to store each bit of data.

- **Advantages:** Faster and more reliable than DRAM.

- **Disadvantages:** More expensive and consumes more power.

- **Example:** Used in cache memory and high-speed registers.

DRAM (Dynamic RAM)

- Description: Stores each bit of data in a separate capacitor.
- Advantages: Higher density and less expensive than SRAM.
- Disadvantages: Slower and requires periodic refreshing to maintain data.
- Example: Used in main memory modules (e.g., DDR4 RAM in computers).

ROM (Read-Only Memory)

- Non-volatile Memory: Retains its contents even when power is turned off.
- Usage: Used for permanent storage of firmware and system software.

PROM (Programmable ROM)

- Description: Can be programmed once after manufacturing.
- Advantages: Customizable by the user.
- Disadvantages: Cannot be reprogrammed.
- Example: Used in embedded systems for storing fixed firmware.

EPROM (Erasable Programmable ROM)

- Description: Can be erased by exposing it to UV light and reprogrammed.
- Advantages: Reusable and reprogrammable.
- Disadvantages: Requires special equipment to erase and reprogram.
- Example: Used in development and testing environments where updates are frequent.

EEPROM (Electrically Erasable Programmable ROM)

- Description: Can be electrically erased and reprogrammed.
- Advantages: Easily reprogrammable without special equipment.
- Disadvantages: Slower write speeds compared to RAM.
- Example: Used in BIOS chips for firmware updates.

Memory Hierarchy and Systems

Memory hierarchy in computer systems is designed to balance speed and cost by using a combination of different types of memory.

Cache Memory

- Description: A small, fast memory located close to the CPU.
- Function: Stores frequently accessed data and instructions to speed up processing.
- Levels: L1 (closest to CPU, smallest, fastest), L2, L3 (larger, slower).
- Example: A modern processor might have 32 KB of L1 cache, 256 KB of L2 cache, and 8 MB of L3 cache.

Main Memory (RAM)

- Description: The primary storage area for data and instructions that the CPU needs while executing tasks.
- Function: Provides a larger, slower storage compared to cache but faster than secondary storage.
- Example: A typical computer might have 8 GB of DDR4 RAM as main memory.

Secondary Storage

- Description: Non-volatile storage for long-term data retention.
- Types: Hard Disk Drives (HDDs), Solid State Drives (SSDs), optical discs.
- Example: An SSD in a laptop provides fast, persistent storage for operating systems, applications, and user data.

Real-world Examples

Smartphones

- RAM: 4-12 GB of DRAM used for running applications and multitasking.
- ROM: Embedded flash memory stores the operating system and application firmware.
- Secondary Storage: Internal SSD provides storage for apps, photos, and files.

Personal Computers

- Cache: Multiple levels of cache (L1, L2, L3) to speed up data access.
- Main Memory: 8-32 GB of DRAM for running the operating system and applications.
- Secondary Storage: HDDs or SSDs for storing software, documents, and media files.

Embedded Systems

- RAM: SRAM for fast, temporary storage.
- ROM: EEPROM or flash memory for storing firmware.
- Secondary Storage: Often minimal or none, as these systems are designed for specific tasks.

16.4 Logic Design

Logic design is the foundation of digital electronics. It involves creating circuits that perform specific logical operations using basic components called logic gates. Logic design is essential for building digital systems such as computers, calculators, and various types of control systems. Understanding how to design and analyze logic circuits is crucial for anyone working in the field of electronics and computer engineering.

Basic Logic Gates (AND, OR, NOT)

Logic gates are the building blocks of digital circuits. Each gate performs a basic logical function.

AND Gate

- **Function:** Outputs true (1) only if all inputs are true.
- **Symbol:** A flat-topped D-shaped symbol.
- **Truth Table:**
  - Inputs: A = 0, B = 0 → Output: 0
  - Inputs: A = 0, B = 1 → Output: 0
  - Inputs: A = 1, B = 0 → Output: 0
  - Inputs: A = 1, B = 1 → Output: 1

Example:

- An AND gate can be used in a security system where both sensors (input) need to be triggered for the alarm (output) to activate.

OR Gate

- **Function:** Outputs true (1) if at least one input is true.
- **Symbol:** A curved D-shaped symbol.

- **Truth Table:**
    - Inputs: A = 0, B = 0 → Output: 0
    - Inputs: A = 0, B = 1 → Output: 1
    - Inputs: A = 1, B = 0 → Output: 1
    - Inputs: A = 1, B = 1 → Output: 1
- **Example:**
- An OR gate can be used in a lighting system where pressing any switch (input) turns on the light (output).

NOT Gate

- **Function:** Outputs the inverse of the input.
- **Symbol:** A triangle with a small circle at the output.
- **Truth Table:**
    - Input: A = 0 → Output: 1
    - Input: A = 1 → Output: 0
- **Example:**
- A NOT gate can be used in an inverter circuit to toggle an LED (input off = output on).

Combinational Logic Circuits

Combinational logic circuits are built using basic logic gates and perform operations based on the current inputs only, without memory of past inputs.

Characteristics

- Output depends solely on the current inputs.
- No storage elements or memory involved.

Example

- **Adder Circuit:** A circuit that adds two binary numbers. A simple half-adder uses an XOR gate for the sum and an AND gate for the carry.

- **Inputs:** A, B

- **Outputs:** Sum, Carry

- **Function:** Sum = A XOR B, Carry = A AND B

Sequential Logic Circuits

Sequential logic circuits differ from combinational circuits in that they include memory elements and their output depends on both current inputs and past inputs (states).

Characteristics

- Output depends on current inputs and previous states.

- Includes storage elements like flip-flops.

Example

Flip-Flop: A basic memory element that stores one bit of data. A D flip-flop captures the value of the input (D) on the rising edge of the clock signal and holds it until the next clock cycle.

- **Inputs:** D (data), CLK (clock)

- **Outputs:** Q (stored data)

- **Function:** On clock edge, Q = D

Example

Counter: A sequential circuit that counts the number of clock pulses. A simple binary counter uses flip-flops to count in binary from 0 to a maximum value.

- **Inputs:** CLK (clock)
- **Outputs:** Binary count value
- **Function:** Increment count on each clock pulse

Real-world Examples

Digital Clocks

- **Combinational Logic:** Used for decoding the time display from binary-coded decimal (BCD) format.
- **Sequential Logic:** Counters and flip-flops track seconds, minutes, and hours.

Computers

- **Combinational Logic:** Used in arithmetic logic units (ALUs) to perform operations like addition, subtraction, and logical comparisons.
- **Sequential Logic:** Registers and memory elements store and process data, maintaining the state of computations.

Traffic Light Controllers

- **Combinational Logic:** Decides the light sequence based on traffic sensors.
- **Sequential Logic:** Timers and state machines manage the timing and transitions between light states.

16.5 Integrated Circuits

Integrated circuits (ICs) are tiny electronic devices made of semiconductor material that contain numerous components such as transistors, resistors, and capacitors. These components are integrated into a single chip to perform various functions in electronic systems. ICs have revolutionized the electronics industry by making it possible to build compact, reliable, and efficient electronic devices.

Types of ICs

Analog ICs

Analog integrated circuits (ICs) are designed to process continuous signals, representing real-world phenomena such as sound, light, and temperature. These ICs are integral to the operation of devices that require amplification, oscillation, and voltage regulation. Analog ICs handle real-world signals by varying voltage or current continuously over time, making them essential in applications like audio equipment, where precise and smooth signal processing is crucial. A prominent example of an analog IC is the operational amplifier (op-amp), widely used in audio systems to amplify sound signals, ensuring clear and high-quality audio output.

Digital ICs

Digital integrated circuits, on the other hand, process discrete signals, making them indispensable in the realm of modern electronics where binary data processing is paramount. These ICs handle binary data, representing information in the form of 0s and 1s, enabling the execution of complex computational tasks. Digital ICs are foundational in constructing logic gates, microprocessors, and memory chips, which are the building blocks of computers, smartphones, and other digital devices. For instance, the microprocessor, a type of digital IC, serves as the brain of a computer, executing instructions and processing data to perform various tasks. Similarly, memory chips in smartphones store and retrieve data efficiently, showcasing the critical role of digital ICs in contemporary technology.

Mixed-Signal ICs

Mixed-signal integrated circuits combine both analog and digital functions on a single chip, offering a versatile solution for applications that require seamless integration of different signal types. These ICs are designed to bridge the gap between analog and digital circuitry, enabling devices to process both continuous and discrete signals effectively. Mixed-signal ICs are essential in various applications, such as audio devices that need to convert analog sound signals to digital data and vice versa. A prime example of mixed-signal ICs is the analog-to-digital converter (ADC) and digital-to-analog converter (DAC). These components are crucial in audio devices, ensuring accurate and high-fidelity conversion between analog and digital signals, thus enhancing the overall performance and functionality of such devices.

Fabrication and Packaging

Fabrication

- **Process:** ICs are fabricated using photolithography, a process where a silicon wafer is patterned with layers of materials to create the desired circuit elements.
- **Steps:**
  - **Doping:** Adding impurities to silicon to create p-type or n-type regions.
  - **Etching:** Removing material to form the circuit patterns.
  - **Deposition:** Adding layers of materials such as metals for interconnections.

Packaging

- **Purpose:** Protects the IC and provides connections to the external circuit.
- **Types:**
  - **DIP (Dual In-line Package):** Pins on both sides, used for through-hole mounting.
  - **SOIC (Small Outline IC):** Smaller size, used for surface mounting.
  - **BGA (Ball Grid Array):** Uses an array of solder balls for connections, suitable for high-density applications.

- **Example:** A microprocessor is fabricated on a silicon wafer using photolithography, then packaged in a BGA package to be mounted on a motherboard.

Applications of Integrated Circuits

Integrated circuits are used in a wide range of applications due to their versatility and efficiency.

Consumer Electronics

- **Usage:** Found in smartphones, tablets, and laptops.
- **Function:** Power management ICs, memory chips, and processors.

Automotive

- **Usage:** Used in engine control units (ECUs), infotainment systems, and safety features.
- **Function:** Control fuel injection, manage infotainment, and enhance safety through sensors and actuators.

Industrial

- **Usage:** Used in automation systems, robotics, and control systems.
- **Function:** Provide precise control and monitoring of industrial processes.

Medical Devices

- **Usage:** Found in diagnostic equipment, wearable health monitors, and implantable devices.
- **Function:** Enable accurate diagnostics and monitoring of patient health.

Real-world Examples

Smartphones

- **Microprocessor IC:** Acts as the brain of the smartphone, executing instructions and managing tasks.
- **Power Management IC:** Manages power distribution, ensuring efficient battery usage.

Automobiles

- **Engine Control Unit (ECU):** Uses ICs to optimize engine performance and fuel efficiency.
- **Airbag System:** ICs process sensor data and deploy airbags in the event of a collision.

Medical Devices

- **Pacemakers:** Use ICs to regulate heartbeats.
- **Glucose Monitors:** ICs process blood glucose levels and provide real-time readings.

ADDITIONAL ONLINE RESOURCES

- **1500+ PRACTICE PROBLEMS [4 PRACTICE EXAMS INCLUDED]**

Enhance your exam preparation with over 1000+ practice problems, complete with step-by-step solutions. Two weeks after your initial download, you will receive a weekly full-length practice exam with step-by-step solutions for four weeks, totaling over 1500 practice problems.

SCAN HERE

Having trouble accessing the resources? Email us at **info@learnovapublishing.com** — we reply within 24 hours, 7 days a week

- **VIDEO RESOURCE RECOMMENDATIONS**

Enhance your learning with our recommended video resources for a deeper understanding of the concepts. When using these resources, please check the current exam specifications.

An abundance of high-quality video resources is available for preparing for this section. For a comprehensive overview, we recommend the following:

YouTube Playlist: *FE Electrical and Computer: Review (Effective from July 2020) by Amit Roy*
Search for this playlist on YouTube. It covers almost all the exam topics, providing detailed explanations and condensing the material effectively.

Online Review Course: For a more organized approach, consider the comprehensive, free, video-based online review course called *"Electrical FE Review"* by Zach Stone, available at electricalfereview.com

Number Systems	
Key Concepts	Handbook Page References
Convert Decimal (Integers) to Binary	
Convert Binary to Decimal (Integers)	
Convert Decimal (Fractions) to Binary	
Convert Binary to Decimal (Fractions)	
Convert Decimal (Integers) to Hexadecimal	
Convert Hexadecimal to Decimal (Integers)	

Boolean Logic	
Key Concepts	Handbook Page References
Idempotent Law	
Associative Law	
Commutative Law	
Distributive Law	
Identity Law	
Complement Law	
DeMorgan's Laws	
Absorption Law	

Logic Gates and Circuits	
Key Concepts	Handbook Page References
Truth Tables of Logic Operators (especially XOR and XNOR)	
Graphical Symbols of Logic Gates	
Logic Minimization	
Key Concepts	Handbook Page References
Sum of Products (SOP)	
Product of Sums (POS)	
Karnaugh Maps (K-Maps)	
Rules for Grouping K-Map Terms for Function Minimization	

Flips-Flops and Counters	
Key Concepts	Handbook Page References
NAND Latch	
NOR Latch	
SR (Set Reset) Flip Flop	
JK Flip Flop	
D Flip Flop	
How Different Flip-Flops and Counters Work	

Programmable Logic Devices and Gate Arrays	
Key Concepts	Handbook Page References
Familiarity with Programmable Logic Devices	

State Machine Design	
Key Concepts	**Handbook Page References**
Navigate between State Diagram and State Table	
Data Path and Controller Design	

Timing	
Key Concepts	**Handbook Page References**
Timing Diagram	
Synchronous and Asynchronous Transmission	
Races and Hazards	

Microprocessors	
Key Concepts	**Handbook Page References**
Computer Architecture	
Components of a Microprocessor:	
- Arithmetic Logic Unit (ALU)	
- Control Unit (CU)	
- Input/Output (I/O)	
- Registers	
- Cache	
- Buses	
Microprocessor Architecture:	
- Harvard Architecture	
- Von Neumann Architecture	
Memory Addressing:	
- Immediate Addressing	
- Register Addressing	

- Direct Addressing	
- Register Indirect Addressing	
- Indexed Addressing	

Memory Technology and Systems	
Key Concepts	**Handbook Page References**
Random Access Memory (RAM)	
Read Only Memory (ROM)	
Cache:	
- Direct Mapped Cache	
- Set Associative Cache	
- Fully Associative Cache	
Replacement Policies:	
- Least Recently Used (LRU)	
- Most Recently Used (MRU)	
- First-In, First-Out (FIFO)	
- Random Replacement Policy	
- Least Frequently Used (LFU)	
Cache Memory Calculations:	
- Tag Bits	
- Index Bits	
- Block Offset Bits	
- Address Bits	

Interfacing
Key Concepts
Interfacing

NOTES/EQUATIONS SUMMARY

DIAGNOSTIC TEST CHECKLIST

Checklist Questions	Yes/No
Have you read and understood all the material in this chapter?	
Are you familiar with all the key concepts and formulas highlighted in the tables for this chapter?	
Have you filled in the tables with the correct NCEES® FE Reference Handbook page numbers for each listed concept and formula?	
Did you download and complete all the designated practice problems for this section?	
Did you review and understand the step-by-step solutions to the practice problems?	
Could you solve the practice problems without excessive reference to the solutions and within a reasonable time frame?	
Are you confident in applying the concepts and formulas from this chapter?	
Have you noted any areas of difficulty or topics that require further review before proceeding?	
Do you know how to quickly find information in the NCEES® FE Reference Handbook?	
Have you reviewed any errors or misunderstandings in this section's material to ensure clarity and understanding?	

CHAPTER 17

Signal Processing

Signal processing is a crucial field in electrical engineering that deals with analyzing, modifying, and synthesizing signals. Signals, which can be in the form of sound, light, or other physical phenomena, are represented mathematically to be processed by various systems. Signal processing techniques are widely used in numerous applications, from telecommunications to biomedical engineering, ensuring that information is accurately transmitted, stored, and interpreted.

In this chapter, we will explore the fundamental concepts of signal processing. We will begin with signal sampling and reconstruction, discussing how continuous signals are converted into discrete forms and then reconstructed. Next, we will delve into time-frequency analysis, explaining the methods used to analyze signals in both time and frequency domains. We will then cover filtering and transformations, essential techniques for modifying and analyzing signals. Finally, we will look at practical applications of signal processing, highlighting how these concepts are applied in real-world scenarios.

17.1 Signal Sampling and Reconstruction

Signal sampling is the process of converting a continuous-time signal into a discrete-time signal by taking periodic samples. This is a fundamental step in digital signal processing, as digital systems can only process discrete data. Sampling allows us to capture and store continuous signals, such as audio or temperature, in a form that can be manipulated by digital devices.

Nyquist-Shannon Sampling Theorem

The Nyquist-Shannon Sampling Theorem provides a criterion for the sampling rate needed to accurately capture a continuous signal without losing information. According to the theorem, the sampling rate must be at least twice the highest frequency present in the signal.

- **Sampling rate** (f_s): The number of samples taken per second.
- **Nyquist rate:** Twice the highest frequency of the signal ($2f_{max}$).

Key Points

- If the signal contains frequencies up to 10 kHz, the minimum sampling rate should be 20 kHz.
- Sampling at a rate lower than the Nyquist rate leads to aliasing.

Example

- For a signal with a maximum frequency of 5 kHz, the minimum sampling rate must be 10 kHz to avoid aliasing.

Aliasing and Anti-aliasing

Aliasing occurs when a signal is sampled at a rate lower than the Nyquist rate, causing different signals to become indistinguishable or "aliased." This results in distortion and loss of information.

Aliasing

The phenomenon where higher frequency components are misinterpreted as lower frequencies during sampling.

Preventing Aliasing

Use an anti-aliasing filter before sampling to remove high-frequency components that could cause aliasing. Ensure the sampling rate is at least twice the highest frequency of the signal.

Example

If an audio signal is sampled at 8 kHz, frequencies above 4 kHz will alias into the 0-4 kHz range, causing distortion. An anti-aliasing filter can remove frequencies above 4 kHz before sampling.

Signal Reconstruction Techniques

Signal reconstruction is the process of converting a discrete-time signal back into a continuous-time signal. This involves interpolating the discrete samples to form a smooth continuous signal.

Reconstruction Methods

- **Zero-Order Hold (ZOH):** Holds each sample value constant until the next sample. Simple but may introduce stair-step artifacts.
- **Linear Interpolation:** Connects sample points with straight lines, providing a smoother approximation than ZOH.
- **Sinc Interpolation:** Uses the sinc function to interpolate between samples, providing the most accurate reconstruction but is computationally intensive.

Example

In digital audio playback, the digital samples are converted back into an analog signal using a digital-to-analog converter (DAC). This often involves applying sinc interpolation to accurately reconstruct the original audio waveform.

Real-world Examples

Digital Audio

- **Sampling:** CDs use a sampling rate of 44.1 kHz to capture audio signals up to 20 kHz, ensuring high-quality sound reproduction without aliasing.
- **Reconstruction:** During playback, the digital samples are converted back into an analog signal by a DAC, providing continuous audio output.

Medical Imaging

- **Sampling:** MRI machines sample signals from the human body at high rates to create detailed images.
- **Reconstruction:** The sampled data is processed and reconstructed into clear, continuous images of internal body structures.

Communication Systems

- **Sampling:** In digital communication, analog signals (e.g., voice) are sampled, digitized, and transmitted over digital networks.
- **Reconstruction:** At the receiver end, the digital signal is reconstructed into an analog signal for playback or further processing.

17.2 Time-Frequency Analysis

Time-frequency analysis is a method used to analyze signals whose frequency content changes over time. This is particularly useful for non-stationary signals, where the signal's characteristics vary over time. Unlike traditional frequency analysis, which provides a global view of the signal's frequency content, time-frequency analysis allows us to observe how the frequency content evolves, providing a more detailed understanding of the signal's behavior.

Fourier Transform

The Fourier Transform (FT) is a mathematical technique used to transform a time-domain signal into its frequency-domain representation. It provides a way to decompose a signal into its constituent frequencies.

- **Function:** Converts a time-domain signal into a spectrum of frequencies.
- **Equation:** $X(f) = \int_{-\infty}^{\infty} x(t) e^{-j2\pi ft} dt$
- **Advantages:** Provides a complete frequency representation of the signal.
- **Disadvantages:** Does not provide time information, only frequency.

Example

Analyzing the frequency content of a recorded audio signal to identify different musical notes.

Key Points

- The FT is useful for stationary signals, where frequency content does not change over time.
- For non-stationary signals, the FT may not provide sufficient information.

Short-Time Fourier Transform (STFT)

The Short-Time Fourier Transform (STFT) is an extension of the Fourier Transform that provides time-localized frequency information. By applying the FT over short, overlapping time windows, STFT captures changes in frequency content over time.

- **Function:** Provides time and frequency information by segmenting the signal into short windows.
- **Equation:** $X(t,f) = \int_{-\infty}^{\infty} (\tau)w(\tau - t)e^{-j2\pi f\tau}d\tau$
    - **w(τ - t):** Window function that isolates a segment of the signal.
- **Advantages:** Captures how frequency content changes over time.
- **Disadvantages:** Trade-off between time and frequency resolution.

Example

Analyzing speech signals to observe how different phonemes (speech sounds) change over time.

Key Points

- The choice of window size affects the resolution: a narrow window provides good time resolution but poor frequency resolution, and vice versa.
- Common window functions include Hamming, Hanning, and Gaussian windows.

Wavelet Transform

The Wavelet Transform (WT) is another time-frequency analysis technique that provides multi-resolution analysis. Unlike the STFT, which uses fixed window sizes, WT uses variable-sized windows to capture both high-frequency details and low-frequency trends.

- **Function:** Analyzes signals at multiple scales using wavelets.
- **Equation:** $W(a,b) = \int_{-\infty}^{\infty} x(t)\psi^*(\frac{t-b}{a})dt$
    - **ψ:** Mother wavelet, **a:** scale parameter, **b:** translation parameter.

- **Advantages:** Provides detailed analysis at multiple resolutions.
- **Disadvantages:** More complex and computationally intensive than STFT.

Example

Analyzing seismic signals to detect different types of waves and events over time.

Key Points

- Wavelets are localized in both time and frequency, providing a flexible approach to signal analysis.
- Common wavelets include the Haar, Daubechies, and Morlet wavelets.

Real-world Examples

Speech Processing

- **STFT:** Used to analyze speech signals for applications such as speech recognition and speaker identification. It helps identify different phonemes and their time-frequency characteristics.
- **Wavelet Transform:** Used to compress and denoise speech signals, improving clarity and reducing storage requirements.

Music Analysis

- **Fourier Transform:** Used to analyze the overall frequency content of a piece of music, identifying dominant notes and harmonics.
- **STFT:** Used to study how the frequency content changes over time, helping in tasks like music transcription and genre classification.

Medical Diagnostics

- **STFT:** Used in electroencephalography (EEG) to study brain waves, identifying different states like sleep stages and epileptic seizures.

- **Wavelet Transform:** Used in electrocardiography (ECG) to analyze heartbeats, detect abnormalities and diagnose heart conditions.

Communications

- **Fourier Transform:** Used in the design and analysis of communication systems, such as modulation and demodulation techniques.
- **Wavelet Transform:** Used in data compression and error detection, improving the efficiency and reliability of communication systems.

17.3 Filtering and Transformations

Filtering is a fundamental signal processing technique used to remove unwanted components from a signal or to extract useful information. Filters can be applied to both analog and digital signals to improve signal quality or to isolate specific frequencies of interest.

Types of Filters

Filters are categorized based on the frequency range they affect. The main types of filters include low-pass, high-pass, band-pass, and band-stop filters.

Low-pass Filter

- **Function:** Allows frequencies below a certain cutoff frequency to pass through and attenuates frequencies above the cutoff.
- **Application:** Used to remove high-frequency noise from audio signals.
- **Example:** A low-pass filter in a speaker system removes high-frequency noise, allowing only the bass sounds to be heard.

High-pass Filter

- **Function:** Allows frequencies above a certain cutoff frequency to pass through and attenuates frequencies below the cutoff.

- **Application:** Used to remove low-frequency noise, such as hums from power lines.
- **Example:** A high-pass filter in a microphone removes low-frequency rumble and hum, improving the clarity of recorded speech.

Band-pass Filter

- **Function:** Allows frequencies within a specific range to pass through and attenuates frequencies outside this range.
- **Application:** Used in radio receivers to select a specific frequency band.
- **Example:** A band-pass filter in a radio tuner isolates the frequency of the desired radio station, filtering out other stations.

Band-stop Filter

- **Function:** Attenuates frequencies within a specific range and allows frequencies outside this range to pass through.
- **Application:** Used to remove a specific frequency, such as a hum at 60 Hz from power lines.
- **Example:** A band-stop filter in audio equipment removes the 60 Hz hum, reducing interference in the audio signal.

Digital Filter Design

Digital filters are implemented using algorithms that process digital signals. The design of digital filters involves selecting the appropriate filter type and defining its characteristics, such as cutoff frequency and filter order.

Design Steps

- **Specification:** Define the filter requirements, including the type (low-pass, high-pass, etc.), cutoff frequency, and acceptable levels of passband and stopband ripple.
- **Design:** Use mathematical methods to design the filter, such as the window method for FIR (Finite Impulse Response) filters or the bilinear transform for IIR (Infinite Impulse Response) filters.
- **Implementation:** Convert the mathematical model into a digital algorithm that can be executed by a digital signal processor (DSP) or software.

Example

Designing a digital low-pass filter to remove high-frequency noise from a digital audio recording involves specifying the cutoff frequency, designing the filter coefficients, and implementing the filter algorithm in software.

Transformations

Transformations are mathematical techniques used to convert signals from one domain to another, facilitating analysis and processing.

Fourier Transform

- **Function:** Converts a time-domain signal into its frequency-domain representation.
- **Application:** Analyzes the frequency content of signals, such as identifying the dominant frequencies in an audio signal.
- **Example:** Using the Fourier Transform to analyze an audio signal reveals the presence of specific musical notes and harmonics.

Laplace Transform

- **Function:** Converts a time-domain signal into the s-domain, used for analyzing linear time-invariant systems.
- **Application:** Analyzes system stability and behavior, such as in control systems.
- **Example:** The Laplace Transform is used to derive the transfer function of a control system, helping engineers design stable and efficient controllers.

Z-Transform

- **Function:** Converts a discrete-time signal into the z-domain, used for analyzing digital systems.
- **Application:** Analyzes the stability and frequency response of digital filters.
- **Example:** The Z-Transform helps design digital filters by analyzing their poles and zeros, ensuring the filters meet specified performance criteria.

Real-world Examples

Audio Processing

- **Filtering:** Low-pass filters remove high-frequency noise from audio recordings, enhancing sound quality.
- **Fourier Transform:** Analyzes the frequency content of recordings to identify and isolate specific sounds.

Communication Systems

- **Filtering:** Band-pass filters isolate the frequency band of interest in a communication signal, reducing interference from other signals.

- **Fourier Transform:** Modulation and demodulation processes convert signals between time and frequency domains for efficient transmission and reception.

Medical Signal Analysis

- **Filtering:** High-pass filters remove low-frequency drift in ECG signals, ensuring accurate heart rate monitoring.
- **Wavelet Transform:** Decomposes EEG signals into different frequency bands to detect and analyze brain wave patterns.

17.4 Practical Signal Processing Applications

Audio Signal Processing

Audio signal processing is a sophisticated field dedicated to manipulating audio signals to enhance sound quality, extract valuable information, or generate new audio effects. This domain employs a variety of techniques to improve recordings, eliminate noise, and create unique audio effects, playing a crucial role in numerous applications such as music production, hearing aids, and speech recognition.

Filtering

In music production, filtering involves the use of low-pass and high-pass filters. Low-pass filters remove high-frequency noise, while high-pass filters eliminate low-frequency hums and rumbles, ensuring a clear and crisp sound.

Equalization

Adjusts the balance between different frequency components of an audio signal. For instance, in music production, boosting bass frequencies can enhance the depth and richness of the sound. Equalization allows sound engineers to sculpt the audio landscape, highlighting desirable frequencies and diminishing unwanted ones.

Compression

Controls the dynamic range of an audio signal, ensuring both loud and soft parts of a track are audible and balanced. This technique amplifies softer sounds and attenuates louder sounds for a uniform output.

Noise Reduction

Identifies and minimizes unwanted noise within an audio signal. This is crucial in various settings, such as during the post-production of recorded interviews. Advanced algorithms and software tools achieve effective noise reduction without compromising the original audio.

Example

In hearing aids, audio signal processing enhances speech clarity and reduces background noise, improving the listening experience for users.

Image Processing

Image processing involves analyzing and manipulating images to improve their quality, extract information, or create visual effects. This field uses various techniques to enhance images, detect patterns, and transform visual data.

Filtering

Filtering techniques can be used to remove noise and enhance the details within an image. A common approach is the application of a Gaussian filter, which smooths the image and reduces noise, leading to clearer and more precise visuals.

Edge Detection

Identifies significant transitions in intensity within an image. This process is essential for recognizing objects and their boundaries. The Sobel operator, for example, is a widely used method for detecting edges, facilitating the identification of objects in various applications, from medical imaging to computer vision systems.

Segmentation

Segmentation is the process of dividing an image into meaningful regions, making it easier to analyze specific parts of the image. In medical imaging, segmentation is used to isolate and examine specific organs or tissues, providing detailed insights that are crucial for accurate diagnosis and treatment.

Compression

Compression techniques reduce the size of image files without significantly degrading their quality. This is particularly important for efficient image storage and transmission. JPEG compression is a commonly used method that balances file size and image quality, making it ideal for storing and sharing photographs and other visual data efficiently.

Example

- In medical imaging, image processing techniques enhance the visibility of anatomical structures in MRI or CT scans, aiding in diagnosis and treatment planning.

Communication Systems

Signal processing in communication systems ensures the reliable transmission and reception of data over various communication channels. Techniques are used to encode, modulate, transmit, receive, and decode signals efficiently.

- **Applications:** Wireless communication, satellite transmission, internet data transfer.

- **Techniques:** Modulation, error correction, filtering, multiplexing.

Modulation

- **Function:** Transforms a signal to fit a specific frequency band for transmission.

- **Example:** Frequency modulation (FM) used in radio broadcasting to encode audio signals.

Error Correction

- **Purpose:** Detects and corrects errors in transmitted data.
- **Example:** Using Hamming codes to correct single-bit errors in digital communication.

Filtering

- **Function:** Removes unwanted frequency components and noise from signals.
- **Example:** Band-pass filters isolate the desired frequency band in wireless communication.

Multiplexing

- **Technique:** Combines multiple signals for transmission over a single channel.
- **Example:** Time-division multiplexing (TDM) in digital telephony allows multiple calls to share the same communication link.

Biomedical Signal Processing

Biomedical signal processing involves analyzing and interpreting biological signals to diagnose and monitor health conditions. Techniques are used to process signals such as ECG, EEG, and EMG for clinical and research applications.

- **Applications:** Cardiology, neurology, patient monitoring.
- **Techniques:** Filtering, feature extraction, pattern recognition, signal averaging.

Filtering

- **Function:** Removes noise and artifacts from biological signals.

- **Example:** High-pass filters remove baseline drift in ECG signals to ensure accurate heart rate measurement.

Feature Extraction

- **Purpose:** Identifies important characteristics of the signal.
- **Example:** Extracting the QRS complex from ECG signals to analyze heart function.

Pattern Recognition

- **Technique:** Identifies specific patterns associated with diseases or conditions.
- **Example:** Detecting seizure activity in EEG signals for epilepsy diagnosis.

Signal Averaging

- **Function:** Reduces noise by averaging multiple instances of a signal.
- **Example:** Averaging evoked potentials in EEG to study brain responses to stimuli.

Real-world Examples

Audio Signal Processing

- **Application:** Digital music production.
- **Example:** Using noise reduction and equalization to produce high-quality audio tracks.

Image Processing

- **Application:** Automated inspection in manufacturing.
- **Example:** Edge detection and segmentation identify defects in products on a production line.

Communication Systems

- **Application:** 4G and 5G wireless networks.
- **Example:** Modulation and error correction techniques ensure reliable data transmission in mobile communication.

Biomedical Signal Processing

- **Application:** Remote patient monitoring.
- **Example:** Filtering and feature extraction from ECG signals provide real-time health data for remote diagnosis and monitoring.

ADDITIONAL ONLINE RESOURCES

- **1500+ PRACTICE PROBLEMS [4 PRACTICE EXAMS INCLUDED]**

Enhance your exam preparation with over 1000+ practice problems, complete with step-by-step solutions. Two weeks after your initial download, you will receive a weekly full-length practice exam with step-by-step solutions for four weeks, totaling over 1500 practice problems.

SCAN HERE

Having trouble accessing the resources? Email us at **info@learnovapublishing.com** — we reply within 24 hours, 7 days a week

- **VIDEO RESOURCE RECOMMENDATIONS**

Enhance your learning with our recommended video resources for a deeper understanding of the concepts. When using these resources, please check the current exam specifications.

An abundance of high-quality video resources is available for preparing for this section. For a comprehensive overview, we recommend the following:

YouTube Playlist: *FE Electrical and Computer: Review (Effective from July 2020) by Amit Roy*
Search for this playlist on YouTube. It covers almost all the exam topics, providing detailed explanations and condensing the material effectively.

Online Review Course: For a more organized approach, consider the comprehensive, free, video-based online review course called *"Electrical FE Review"* by Zach Stone, available at electricalfereview.com

Sampling	
Key Concepts	**Handbook Page References**
Nyquist Sampling Theorem	
Aliasing	

Analog and Digital Filters	
Key Concepts	**Handbook Page References**
Analog Filters:	
- First Order Low Pass and High Pass Filters	
- Band-Pass Filter	
- Band-Stop Filter	
Digital Filters	

Z-Transforms and Convolution	
Key Concepts	**Handbook Page References**
Z-Transforms:	
- Z-Transforms Pairs:	
- *Delta*	
- *Unit Ramp*	
- *Unit Step*	
- *Exponential*	
- *Sinusoidal*	
- Z-Transform Properties:	
- *Linearity*	
- *Time Shifting*	
- *Time Scaling*	

- *Time Reversal*	
- *Convolution*	
- *Time Difference*	
- *Z-Domain Differentiation*	
Convolution:	
- Continuous Time Convolution:	
- *Graphical Convolution*	
- Discrete Time Convolution	
- Convolution Properties:	
- *Identity (Delta Function)*	
- *Shift Property*	
- *Commutative*	
- *Associative*	
- *Distributive*	

NOTES/EQUATIONS SUMMARY

DIAGNOSTIC TEST CHECKLIST

Checklist Questions	Yes/No
Have you read and understood all the material in this chapter?	
Are you familiar with all the key concepts and formulas highlighted in the tables for this chapter?	
Have you filled in the tables with the correct NCEES® FE Reference Handbook page numbers for each listed concept and formula?	
Did you download and complete all the designated practice problems for this section?	
Did you review and understand the step-by-step solutions to the practice problems?	
Could you solve the practice problems without excessive reference to the solutions and within a reasonable time frame?	
Are you confident in applying the concepts and formulas from this chapter?	
Have you noted any areas of difficulty or topics that require further review before proceeding?	
Do you know how to quickly find information in the NCEES® FE Reference Handbook?	
Have you reviewed any errors or misunderstandings in this section's material to ensure clarity and understanding?	

CHAPTER 18

Software Development

Software development is a crucial aspect of modern technology, underpinning everything from simple applications to complex systems used in various industries. Understanding the principles of software development enables engineers to create efficient, reliable, and maintainable software solutions. This chapter will cover the essential topics of algorithms and data structures, programming fundamentals, software design and testing, and software project management.

18.1 Algorithms and Data Structures

Algorithms and data structures are the backbone of computer science and software development. An algorithm is a step-by-step procedure for solving a problem or performing a task, while a data structure is a way of organizing and storing data to facilitate efficient access and modification. Together, they form the foundation for creating efficient and effective software.

- **Algorithms:** Define the logic for solving specific problems.
- **Data Structures:** Provide ways to store and organize data.

Understanding these concepts is crucial for developing software that is not only correct but also performs well, particularly when dealing with large amounts of data or complex operations.

Algorithms can be categorized based on the type of problems they solve. Two of the most fundamental types are sorting and searching algorithms.

Sorting Algorithms

Sorting algorithms arrange data in a specific order (e.g., ascending or descending). Two common sorting algorithms are Bubble Sort and Quick Sort.

Bubble Sort

A simple comparison-based algorithm that repeatedly steps through the list, compares adjacent elements and swaps them if they are in the wrong order.

Complexity: Bubble Sort has a time complexity of $O(n^2)$ in both average and worst-case scenarios. This quadratic complexity makes it inefficient for large lists when compared to more advanced sorting algorithms like Quick Sort or Merge Sort.

Steps

1. Begin with the first element in thc list and compare it with the next element.
2. If the first element is greater than the second, swap them. This ensures that the larger element moves towards the end of the list.
3. Continue comparing and swapping adjacent elements for the entire length of the list.
4. After each complete pass through the list, the next largest element will be in its correct position.
5. Repeat the steps for all elements, gradually reducing the range of comparison as the largest elements settle into their correct positions at the end of the list.

6. Continue the process until no more swaps are needed, indicating that the list is sorted.

Example: Sorting the list [5, 3, 8, 4, 2] using Bubble Sort:

Pass 1: [3, 5, 4, 2, 8]

Pass 2: [3, 4, 2, 5, 8]

Pass 3: [3, 2, 4, 5, 8]

Pass 4: [2, 3, 4, 5, 8]

Quick Sort

An efficient divide-and-conquer algorithm that selects a 'pivot' element and partitions the array into two sub-arrays according to whether elements are less than or greater than the pivot.

Complexity

Quick Sort has an average time complexity of O (n log n), making it much faster on average than simpler algorithms like Bubble Sort or Insertion Sort. However, its worst-case time complexity is $O(n^2)$, which can occur when the pivot selection is poor (e.g., always selecting the smallest or largest element in the sub-array). This worst-case scenario can often be mitigated by using strategies such as randomized pivot selection or the median-of-three method.

Steps

1. Choose a pivot element.
2. Partition the array into two sub-arrays.
3. Recursively apply Quick Sort to the sub-arrays.

Example: Sorting the list [5, 3, 8, 4, 2] using Quick Sort:

Pivot: 5

Partition: [3, 4, 2], 5, [8]

Recursively sort sub-arrays: [2, 3, 4], 5, [8]

Searching Algorithms

Searching algorithms are used to find specific elements within a data structure. Two common searching algorithms are Linear Search and Binary Search.

Linear Search

A simple search algorithm that checks each element in the list sequentially until the desired element is found or the list is exhausted.

Complexity: O(n)

Steps

1. Start from the first element.
2. Compare each element with the target value.
3. Return the index if the target is found, or -1 if not found.

Example: Searching for 4 in the list [5, 3, 8, 4, 2]:

- Compare with 5: no match.
- Compare with 3: no match.
- Compare with 8: no match.
- Compare with 4: match found at index 3.

Binary Search

An efficient search algorithm for sorted lists that repeatedly divides the search interval in half.

Complexity: O(log n)

Steps

1. Start with the middle element.

2. If the target is equal to the middle element, return the index.
3. If the target is less than the middle element, repeat the search on the left sub-array.
4. If the target is greater, repeat on the right sub-array.

Example: Searching for 4 in the sorted list [2, 3, 4, 5, 8]:

- Compare with 4 (middle element): match found at index 2.

Essential Data Structures

Data structures provide efficient ways to store and organize data, enabling faster access and manipulation. Understanding the characteristics, advantages, and disadvantages of various data structures is essential for developing optimized algorithms and software applications.

Arrays

An array is a collection of elements identified by index or key. It allows for fast access to elements by their index, making it efficient for retrieving and updating values. However, arrays have a fixed size, which means that once an array is created, its size cannot be changed. Additionally, inserting or deleting elements can be costly because it may require shifting other elements to maintain the order.

Example: Storing the list of numbers [5,3,8,4,2] in an array can be done as follows:

arr[0] = 5

arr[1] = 3

arr[2] = 8

arr[3] = 4

arr[4] = 2

Linked Lists

A linked list is a linear collection of elements, where each element points to the next. This structure allows for dynamic sizing, meaning it can grow or shrink as needed. Linked lists provide efficient insertion and deletion operations because only the links between nodes need to be updated. However, accessing elements is slower compared to arrays because it requires traversing the list from the beginning to the desired position.

Example: Representing the list [5,3,8,4,2] in a linked list involves creating nodes where each node contains a value and a reference to the next node. For instance, the first node contains the value 5 and a reference to the second node, which contains the value 3, and so on.

Stacks and Queues

Stacks

A stack is a collection of elements with Last In, First Out (LIFO) access. Operations include:

- Push: Add an element to the top of the stack.
- Pop: Remove the element from the top of the stack.
- Peek: View the element at the top of the stack without removing it.

Queues

A queue is a collection of elements with First In, First Out (FIFO) access. Operations include:

- **Enqueue:** Add an element to the end of the queue.
- **Dequeue:** Remove the element from the front of the queue.
- **Peek:** View the element at the front of the queue without removing it.

Example: Stacks are commonly used for undo operations in text editors, where the most recent action is the first to be undone. Queues are used for task scheduling in operating systems, ensuring tasks are processed in the order they arrive.

Trees and Graphs

Trees

A tree is a hierarchical data structure with a root node and child nodes. Types of trees include:

- **Binary Trees:** Each node has at most two children.
- **AVL Trees:** A self-balancing binary search tree where the difference in heights between the left and right subtrees is at most one.
- **B-trees:** A balanced tree data structure that maintains sorted data and allows for efficient insertion, deletion, and search operations.

Graphs

A graph is a collection of nodes (vertices) connected by edges. Types of graphs include:

- **Directed Graphs:** Edges have a direction, indicating the relationship flows from one vertex to another.
- **Undirected Graphs:** Edges have no direction, indicating a bi-directional relationship.
- **Weighted Graphs:** Edges have weights, representing the cost or distance between vertices.
- **Unweighted Graphs:** Edges have no weights.

Trees are used in file systems to represent directories and files hierarchically. Graphs are used to represent networks, such as social networks or computer networks, where nodes represent entities and edges represent connections.

Real-world Examples

Algorithms in Real-world Applications

- **Sorting Algorithms:** Quick Sort is widely used in programming libraries due to its efficiency in sorting large datasets. Its average-case performance of O (n log n) makes it suitable for many applications, including database sorting and search algorithms.

- **Searching Algorithms:** Binary Search is employed in search engines and databases to quickly locate data in sorted structures. With a time complexity of O (log n), it is highly efficient for large datasets.

Data Structures in Real-world Applications

- **Arrays:** Arrays are used in applications requiring quick access to elements, such as image processing, where each pixel can be accessed directly by its index.

- **Linked Lists:** Linked lists are employed in dynamic data structures like memory allocation systems, where efficient insertion and deletion of elements are crucial.

- **Stacks and Queues:** Stacks are used in compiler design for syntax parsing, where elements are processed in reverse order of their addition. Queues are essential in operating systems for process scheduling, ensuring tasks are executed in the order they arrive.

- **Trees and Graphs:** Trees are crucial for hierarchical data representation, such as organizational structures or XML parsing. Graphs are indispensable for network topology representations, used in routing algorithms, social network analysis, and other applications involving complex relationships between entities.

18.2 Programming Fundamentals

Programming is the process of creating instructions that a computer can follow to perform specific tasks. These instructions, or code, are written in programming languages that provide the syntax and structure for communicating with the computer. Learning to program is essential for developing software, automating tasks, and solving complex problems in various domains.

Basic Programming Concepts

Understanding the basic concepts of programming is crucial for writing efficient and effective code.

Variables and Data Types

Variables are used to store data that can be manipulated by the program. Data types define the kind of data a variable can hold.

- **Variables:** Containers for storing data values.
    - **Example:** `int age = 25;` where `age` is a variable storing the integer value 25.
- **Data Types:** Define the type of data a variable can hold.
    - **Common Data Types**
        - **Integer (int):** Whole numbers (e.g., 1, 2, 3)
        - **Float (float):** Decimal numbers (e.g., 1.5, 3.14)
        - **Character (char):** Single characters (e.g., 'a', 'b')
        - **String (string):** Sequence of characters (e.g., "hello")
        - **Boolean (bool):** True or false values (e.g., true, false)

Control Structures

Control structures are used to determine the flow of execution in a program. They include conditional statements and loops.

- **If-else Statements**
  - **Purpose:** Execute code blocks based on conditions.
  - **Example**

```
if (age >= 18) {
   cout << "Adult";
} else {
   cout << "Minor";
}
```

- **Loops**
  - **Purpose:** Repeat code blocks multiple times.
  - **Types**
    - **For Loop:** Repeats a block of code a specified number of times.
      - **Example:**

```
for (int i = 0; i < 5; i++) {
    cout << i;
}
```

    - **While Loop:** Repeats a block of code as long as a condition is true.
  - **Example**

```
int i = 0;
while (i < 5) {
   cout << i;
   i++;
}
```

Functions and Procedures

Functions and procedures are reusable blocks of code that perform specific tasks.

- **Functions**
  - **Purpose:** Perform tasks and return a value.
  - **Syntax**

    ```
    int add(int a, int b) {
       return a + b;
    }
    ```

  - **Example:** Calling add(3, 4) returns 7.
- **Procedures:**
  - **Purpose:** Perform tasks without returning a value (void functions).
  - **Syntax**

    ```
    void printMessage() {
       cout << "Hello, World!";
    }
    ```

  - **Example:** Calling printMessage() prints "Hello, World!".

Object-Oriented Programming (OOP) Basics

Object-Oriented Programming (OOP) is a programming paradigm that uses objects to represent real-world entities. It enhances code reusability and modularity.

Classes and Objects

- **Classes:** Blueprints for creating objects. Define properties (attributes) and behaviors (methods).
  - **Example**

    ```
    class Car {
    public:
       string brand;
       int year;
    ```

```
    void start() {
        cout << "Car started";
    }
};
```

- **Objects:** Instances of classes.
  - **Example**

```
Car myCar;
myCar.brand = "Toyota";
myCar.year = 2020;
myCar.start();
```

Inheritance and Polymorphism

- **Inheritance:**
  - **Purpose:** Allows a class to inherit properties and methods from another class.
  - **Example**

```
class Vehicle {
public:
    string brand;
    void honk() {
        cout << "Honk!";
    }
};

class Car : public Vehicle {
public:
    int year;
```

```
};

Car myCar;
myCar.brand = "Toyota";
myCar.honk(); // Output: Honk!
```

- **Polymorphism**
  - **Purpose:** Allows methods to have different implementations based on the object that invokes them.
  - **Example**

```
class Animal {
public:
   virtual void makeSound() {
      cout << "Some sound";
   }
};

class Dog : public Animal {
public:
   void makeSound() override {
      cout << "Bark";
   }
};

Animal* myDog = new Dog();
myDog->makeSound(); // Output: Bark
```

18.3 Software Design and Testing

Software design is a critical phase in the software development lifecycle that involves creating a blueprint for the construction of a software system. The design process defines the architecture, components, interfaces, and data for a system to satisfy specified requirements. Good software design ensures that a system is robust, scalable, maintainable, and meets user needs.

- **Purpose:** Provide a clear plan for building a software system.
- **Benefits:** Enhances code quality, facilitates maintenance, and reduces development costs.

Design Principles

Adhering to design principles helps in creating a well-structured and efficient software system.

Modular Design

Modular design involves dividing a software system into separate, interchangeable modules, each responsible for a specific functionality. This approach improves maintainability, scalability, and reusability.

- **Characteristics**
    - **Encapsulation:** Each module encapsulates its data and functions.
    - **Separation of Concerns:** Different functionalities are separated into distinct modules.
- **Example:** In a web application, the frontend, backend, and database are designed as separate modules, each handling specific tasks.

Design Patterns

Design patterns are proven solutions to common software design problems. They provide templates for solving recurring design issues and promote best practices.

Singleton Pattern

- **Purpose:** Ensure a class has only one instance and provide a global point of access to it.
- **Usage:** Managing a connection pool, logging, configuration settings.
- **Example**

```
class Singleton {
private:
    static Singleton* instance;
    Singleton() {}
public:
    static Singleton* getInstance() {
        if (!instance)
            instance = new Singleton();
        return instance;
    }
};
Singleton* Singleton::instance = nullptr;
```

Observer Pattern

- **Purpose:** Define a one-to-many dependency between objects, so that when one object changes state, all its dependents are notified and updated automatically.
- **Usage:** Event handling systems, data binding in GUI frameworks.

- **Example**

```
class Observer {
public:
    virtual void update() = 0;
};

class Subject {
private:
    list<Observer*> observers;
public:
    void attach(Observer* observer) {
        observers.push_back(observer);
    }
    void notify() {
        for (Observer* observer : observers)
            observer->update();
    }
};
```

Software Testing

Software testing is the process of evaluating a software system to detect differences between given input and expected output. It ensures the software meets the required standards and works as intended.

Unit Testing

- **Description:** Tests individual components or modules of a software system to ensure they work correctly in isolation.
- **Purpose:** Identify and fix bugs at an early stage.

- **Tools:** JUnit, NUnit, TestNG.
- **Example**

```
public void testAddition() {
    Calculator calc = new Calculator();
    assertEquals(5, calc.add(2, 3));
}
```

Integration Testing

- **Description:** Tests the interaction between integrated modules to ensure they work together as expected.
- **Purpose:** Detect issues in the interaction between modules.
- **Tools:** JUnit, TestNG, Selenium.
- **Example**

```
public void testIntegration() {
    UserService userService = new UserService();
    AuthService authService = new AuthService();
    userService.setAuthService(authService);
    assertTrue(userService.authenticate("user", "pass"));
}
```

Testing Techniques and Tools

Black-box Testing

- **Description:** Testing without knowledge of the internal workings of the system.
- **Purpose:** Focus on input-output behavior.
- **Tools:** Selenium, QTP.

White-box Testing

- **Description:** Testing with knowledge of the internal code structure.
- **Purpose:** Focus on testing internal operations and code paths.
- **Tools:** JUnit, NUnit.

Automated Testing

- **Description:** Using tools to automate the execution of tests.
- **Purpose:** Increase efficiency and coverage, reduce manual effort.
- **Tools:** Selenium, Jenkins, JUnit.

18.4 Software Project Management

Software project management involves planning, organizing, and overseeing the development and delivery of software projects. Effective project management ensures that software is delivered on time, within budget, and meets the required quality standards. It encompasses various activities, including defining project goals, managing resources, and mitigating risks.

Project Lifecycle

The project lifecycle consists of several phases that guide the development process from start to finish.

Planning and Requirements Gathering

Define project goals, scope, and deliverables. Gather detailed requirements from stakeholders to ensure the final product meets their needs.

Activities:

- Conducting stakeholder interviews and workshops.
- Creating a project plan with timelines, resources, and budget.

- Developing a requirements specification document.

Example: For an online shopping platform, requirements gathering would involve understanding the features needed, such as product listings, shopping cart, and payment gateway integration.

Design and Implementation

Create detailed designs based on the gathered requirements and implement the software.

- **Activities**
  - Designing system architecture and user interfaces.
  - Writing code and integrating various components.
  - Conducting code reviews and unit testing.

Example: Designing a user-friendly interface for a banking app and implementing backend services for transaction processing.

Testing and Deployment

Ensure the software is bug-free and performs as expected. Deploy the software to the production environment.

- **Activities**
  - Performing various types of testing (unit, integration, system, acceptance).
  - Fixing bugs and optimizing performance.
  - Deploying the software and conducting post-deployment checks.

Example: Testing a healthcare management system to ensure it handles patient data correctly and securely before deploying it to hospitals.

Project Management Methodologies

Different methodologies guide the software development process, each with its own approach and principles.

Agile

An iterative and incremental approach that emphasizes flexibility, collaboration, and customer feedback.

- **Characteristics**
    - Short development cycles called sprints.
    - Regular communication with stakeholders.
    - Continuous delivery of small, functional software increments.

Example: Using Agile methodology to develop a social media app, with new features added and tested in two-week sprints.

Waterfall

A linear and sequential approach where each phase depends on the deliverables of the previous one.

- **Characteristics**
    - Clearly defined stages (requirements, design, implementation, testing, deployment).
    - Emphasis on thorough documentation.
    - Minimal changes once a phase is completed.

Example: Using the Waterfall model to develop critical systems like air traffic control software, where changes are costly and require extensive testing.

ADDITIONAL ONLINE RESOURCES

- **1500+ PRACTICE PROBLEMS [4 PRACTICE EXAMS INCLUDED]**

Enhance your exam preparation with over 1000+ practice problems, complete with step-by-step solutions. Two weeks after your initial download, you will receive a weekly full-length practice exam with step-by-step solutions for four weeks, totaling over 1500 practice problems.

SCAN HERE

Having trouble accessing the resources? Email us at **info@learnovapublishing.com** — we reply within 24 hours, 7 days a week

- **VIDEO RESOURCE RECOMMENDATIONS**

Enhance your learning with our recommended video resources for a deeper understanding of the concepts. When using these resources, please check the current exam specifications.

For this section, we recommend the following:

Online Review Course: Consider the comprehensive, free, video-based online review course called *"Electrical FE Review"* by Zach Stone, available at electricalfereview.com

Algorithms	
Key Concepts	**Handbook Page References**
Sorting Algorithms:	
- Bubble Sort	
- Merge Sort	
- Heap Sort	
- Quick Sort	
Searching Algorithms:	
- Binary Search	
- Hashing	
Algorithm Efficiency:	
- Big-O Notation	
Tree Traversal:	
- In-Order Traversal	
- Pre-Order Traversal	
- Post-Order Traversal	
Graph Traversal:	
- Depth-First Search (DFS)	
- Breadth-First Search (BFS)	

Data Structures	
Key Concepts	**Handbook Page References**
Types of Data Structures:	
- Static (Arrays, Structures)	
- Dynamic (Linked Lists, Trees, Graphs)	
Object-Oriented Programming Concepts:	
- Class and Object Definitions	

- Inheritance, Polymorphism, Encapsulation	

Software Implementation	
Key Concepts	**Handbook Page References**
Software Design Methods:	
- Waterfall	
- Iterative	
- V-Model	
Software Syntax and Flow Charts:	
- Flow Chart Symbols and Development	
Unified Modelling Language (UML):	
- Extends	
- Implements	
- Class Relationships	
Software Testing:	
- McCabe's Cyclomatic Complexity	

NOTES/EQUATIONS SUMMARY

DIAGNOSTIC TEST CHECKLIST

Checklist Questions	Yes/No
Have you read and understood all the material in this chapter?	
Are you familiar with all the key concepts and formulas highlighted in the tables for this chapter?	
Have you filled in the tables with the correct NCEES® FE Reference Handbook page numbers for each listed concept and formula?	
Did you download and complete all the designated practice problems for this section?	
Did you review and understand the step-by-step solutions to the practice problems?	
Could you solve the practice problems without excessive reference to the solutions and within a reasonable time frame?	
Are you confident in applying the concepts and formulas from this chapter?	
Have you noted any areas of difficulty or topics that require further review before proceeding?	
Do you know how to quickly find information in the NCEES® FE Reference Handbook?	
Have you reviewed any errors or misunderstandings in this section's material to ensure clarity and understanding?	

CHAPTER 19

Online Resources

Welcome to the final chapter of your study guide, where we have consolidated all the essential online resources to support your FE Electrical and Computer exam preparation. Throughout this book, you have encountered various links and QR codes providing access to valuable materials such as practice problems, mock exams, and video resource recommendations. Here, you can conveniently find these resources in one place.

Access Your Online Resources:

-1500+ Practice Problems [4 Practice Exams Included]

Enhance your exam preparation with over 1000+ practice problems, complete with step-by-step solutions. Two weeks after your initial download, you will receive a weekly full-length practice exam with step-by-step solutions for four weeks, totaling over 1500 practice problems.

SCAN HERE

-For the **video resource recommendations**, please refer to the end of each review chapter.

If you encounter any issues accessing these resources, find any errors, have suggestions for improvements, or have any other questions, please do not hesitate to email us at **info@learnovapublishing.com**. We are here to support you and ensure you have everything you need to succeed.

Closing Remarks

As you reach the end of this study guide, we want to congratulate you on your dedication and hard work. Preparing for the FE Electrical and Computer exam is a significant step toward achieving your professional goals. We wish you the best of luck on your exam day and in your future engineering career. Remember, perseverance and a positive mindset are key to your success. You've got this!

Best wishes,

The Learnova Team

Made in United States
North Haven, CT
10 November 2026

22088261R00187